GRAND COMPLICATIONS

VOL. XI

THE ORIGINAL ANNUAL OF THE WORLD'S WATCH COMPLICATIONS AND MANUFACTURERS® WATCHES

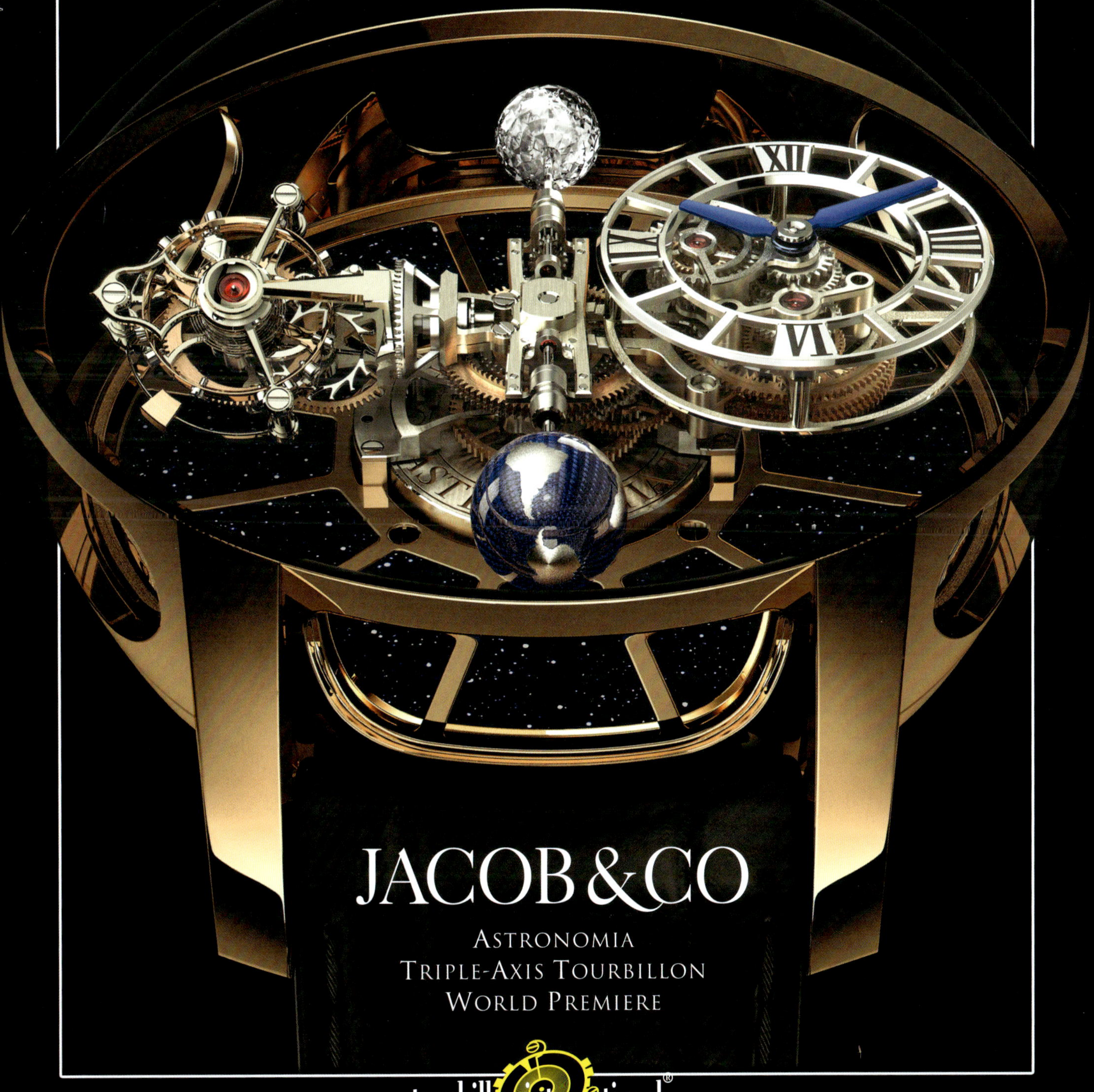

tourbillon international®

GRAND COMPLICATIONS®

THE ORIGINAL ANNUAL OF THE WORLD'S WATCH COMPLICATIONS AND MANUFACTURERS®

SPECIAL ASTRONOMICAL WATCH EDITION

First published in the United States in 2015 by

TOURBILLON INTERNATIONAL
A MODERN LUXURY COMPANY

11 West 25th Street, 8th Floor
New York, NY 10010
Tel: +1 (212) 627-7732 Fax +1 (312) 274-8418
www.modernluxury.com/watches

In association with RIZZOLI INTERNATIONAL PUBLICATIONS, INC.

300 Park Avenue South, New York, NY 10010

ISBN: 978-0-8478-4555-2

PRINTED IN ITALY

COVER: ASTRONOMIA (JACOB & CO.)

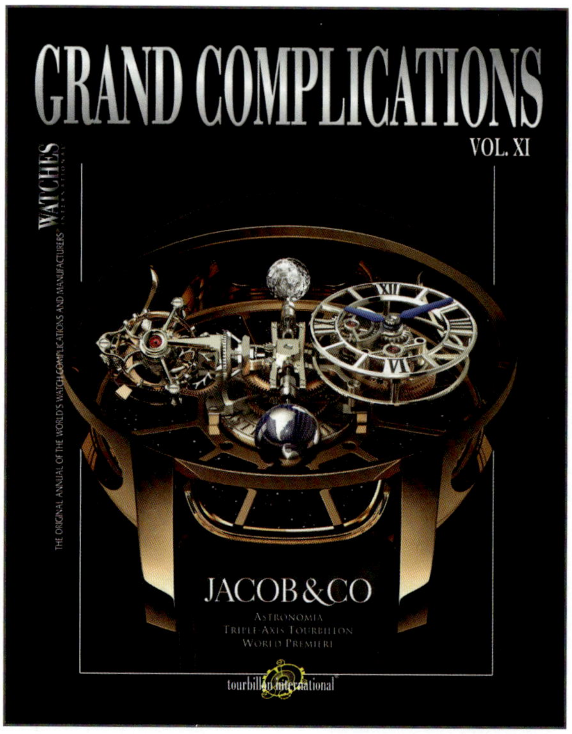

Breguet
Depuis 1775
Breguet, the innovator.
Invention of the shock-protection device, 1790
Inspired by "subscription watches", the Tradition 7027BR model daringly symbolizes the Breguet art of watchmaking through a subtle play on transparency effects and an eminent contemporary architectural design. It highlights one of Breguet's most important inventions, the *pare-chute*, designed to protect the balance pivots in case of impact, it was the forerunner of all modern shock-absorbing devices. History is still being written...
arije

CALIBER EXTRA FLAT
TOURBILLON RM 017

Letter from the President

Daring to go further than ever before

Like a salon for the most daring of minds, *Grand Complications* brings together haute horology's exceptional visionaries and their intricate timepieces.

Vibrant masterpieces of mechanical complexity, the watches presented in this book take timekeeping to new heights, seeking new and exhilarating ways to manifest the technical genius contained in calibers smaller than the palm of one's hand. These creations, containing hundreds of minuscule components perfectly arranged to work as one symphony of motion, demonstrate the result of a meeting between tradition and an unquenchable thirst for innovation. They are what happens when the world's foremost inventors challenge themselves to take on the previously unfathomable.

Once used to indicate the hours and minutes, mechanical watches bring to the measurement of time a texture and depth like never before. From chronographs capable of measuring elapsed intervals of small fractions of a second, to diving companions whose genius emerges when they plunge to the depths of the ocean, to tourbillons that foil gravity's most valiant efforts to challenge horological precision, to timepieces fit to take their owners to all corners of the earth with their ingenious incorporations of multiple time zones, to movements calibrated to respond seamlessly to the calendar's most capricious irregularities, *Grand Complications* sheds a light on time kept with magnificent complexity.

Whether brought to life in sound by the resonating strike of a hammer, or indicated in ways that reject conventional visual circularity with retrograde modules, even our ways of experiencing the passing of time are delightfully challenged.

These timepieces represent the finest unions of finesse, ingenuity and micro-mechanical excellence, answering the question of whether we dare with a resounding "yes!"

Michael Dickey

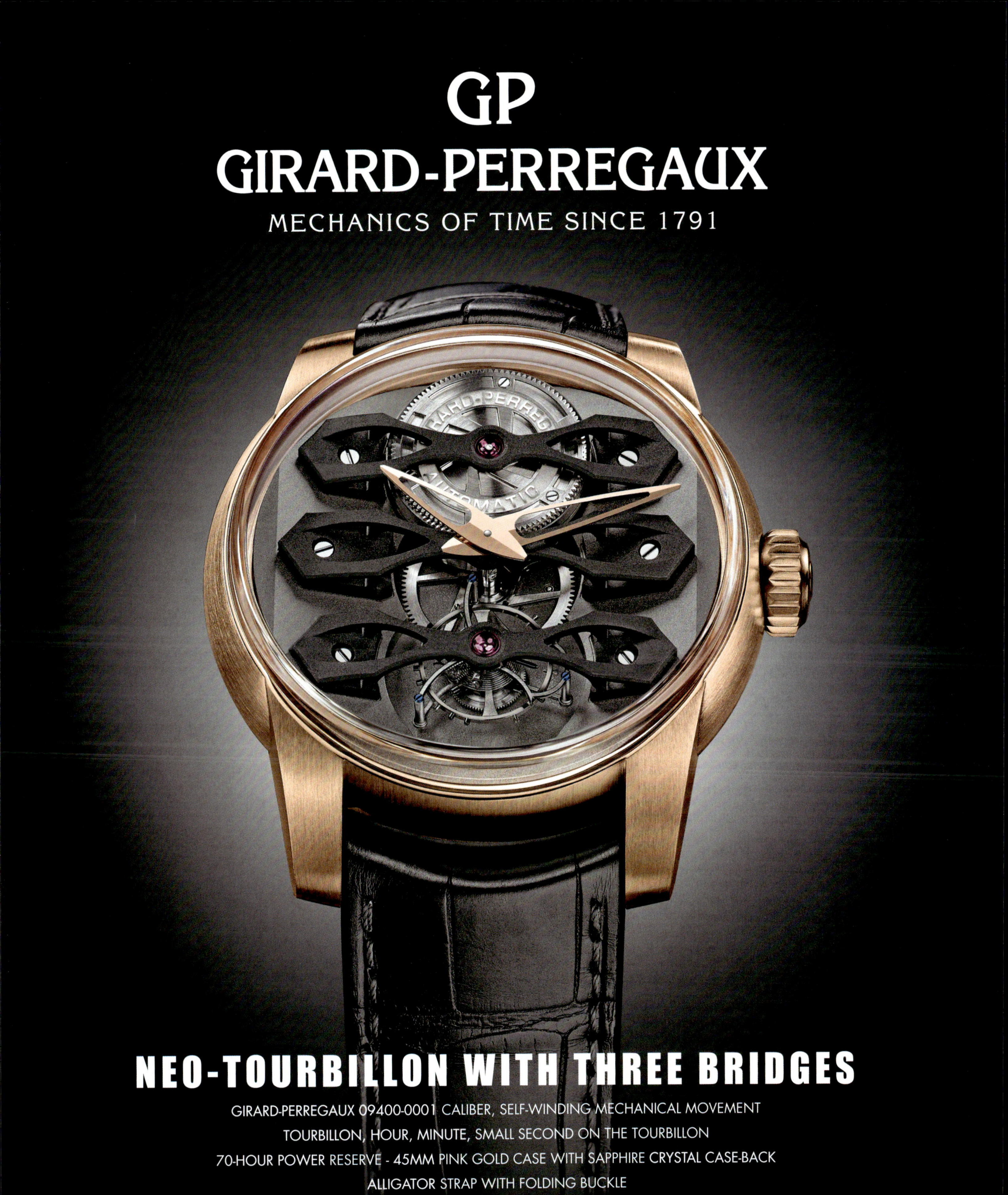
GP
GIRARD-PERREGAUX
MECHANICS OF TIME SINCE 1791
NEO-TOURBILLON WITH THREE BRIDGES
GIRARD-PERREGAUX 09400-0001 CALIBER, SELF-WINDING MECHANICAL MOVEMENT
TOURBILLON, HOUR, MINUTE, SMALL SECOND ON THE TOURBILLON
70-HOUR POWER RESERVE - 45MM PINK GOLD CASE WITH SAPPHIRE CRYSTAL CASE-BACK
ALLIGATOR STRAP WITH FOLDING BUCKLE

cellini

Whether you're a seasoned collector or thinking about purchasing your first fine timepiece, Cellini Jewelers will tempt you with a carefully curated collection of watches from the world's finest brands.

Widely regarded as an horological tastemaker for nearly 40 years, Cellini's flair for breaking international brands into the American market is unparalleled. Both of its Manhattan boutiques were early to carry watches by A. Lange & Söhne, Greubel Forsey, Franck Muller and Hublot long before their popularity soared among collectors. Today the independent jeweler continues to be an influential launching pad for brands like De Bethune, HYT and Ludovic Ballouard.

Last year, Cellini added no less than four brands: Arnold & Son, Clerc, Urwerk and Waltham. The latter is the extraordinary American brand that helped revolutionize the watch industry in the 19th century. Re-launched in 2014, Waltham's Swiss-made timepieces are sporty interpretations of the brand's historic aeronautical clocks, including the cockpit timekeeper that accompanied Charles Lindbergh as he completed the first non-stop transatlantic flight in 1927.

For its modern re-launch, Waltham unveiled the Aeronaval collection, which consists of three models: Waltham XA, Waltham CDI and Waltham ETC. "The history of Waltham is truly incredible and not really well known. I was blown away when I started learning more about what they've done and where they've been—from its precise railroad pocket watches to its expeditions to the poles," says Cellini President Leon Adams.

Whether the brand is just starting out like Waltham, or an instantly recognizable icon like Jaeger-LeCoultre, Adams considers it a point of pride to carry a company's entire collection. That incredible depth creates an unparalleled experience that resonates deeply with people who are passionate about watches. What Cellini does best, Adams says, is use that range to help someone find the right watch. He explains, "If you like a particular complication or style, we line up models from several brands so you can weigh your options and judge for yourself what looks and feels right. You can't find that anywhere else, especially a mono-brand boutique."

▲ **BLACK GOLD OPAL EARRINGS**
Pink spinel adds a modern flair to these eye-catching opal and brown diamond drop earrings.

▲ **BLACK GOLD OPAL RING**
Black gold mirrors the antique feel of this opal ring with brown diamonds.

◀ **WALTHAM XA**
The Waltham XA takes its design cues from the clock Charles Lindbergh used in the Spirit of St. Louis.

MADISON AVENUE BOUTIQUE

HOTEL WALDORF-ASTORIA BOUTIQUE

SPARKLING PERSONALITY

But Cellini is known for more than just watches. Its boutiques are also ranked among Manhattan's finest jewelers. The company's impeccably high standards are reflected in everything from unrivaled selection and superior quality to the attentive experts who are ready to guide you through Cellini's glamorous universe.

From classic to contemporary, and subtle to stunning, the jewelry at Cellini encompasses a wide range of styles. Within this exquisite exhibition of glittering luxury, you'll find diamonds in every hue, impressive strands of pearls and magnificent emeralds, rubies and sapphires in one-of-a-kind, handmade settings. "We strive to create a world where the only limit is your imagination," Adams says.

To stay ahead of changing tastes, the collection evolves constantly, adding rare and unique pieces from around the world. For instance, as the popularity of black gold has exploded, Cellini has stayed at the forefront by seeking out the most innovative the designs. The wide cuff bracelet and opal jewelry featured here underscore black gold's inexhaustible versatility.

Cellini also introduced a range of jewelry creations that provide a colorful counterpoint to black gold's dark beauty. Each piece plays with perception by using transparent gemstones to add striking depth to the intricate patterns below. Built from the bottom up, most of the elaborate metalwork and gem setting is topped by faceted quartz, amethyst or topaz for a look that is guaranteed to spark its share of conversations.

Whether it's fine jewelry or collectible timepieces, Cellini is the place to see what's next.

DOUBLE TOURBILLON 30° TECHNIQUE BI-COLOR
(Greubel Forsey)

HYDROSCAPH H1 CHRONOMETER
(Clerc)

DB28 MAXICHRONO
(De Bethune)

Hotel Waldorf-Astoria
301 Park Avenue at 50th Street
New York • NY 10022
212-751-9824

509 Madison Avenue at 53rd Street
New York • NY 10022
212-888-0505

800-CELLINI
www.CelliniJewelers.com

▲ **PINK AMETHYST NECKLACE**

White diamonds, pink sapphires and pink tourmalines are set underneath faceted pink amethyst, crafted in 18-karat rose gold.

▶ **BLUE TOPAZ EARRINGS**

Faceted blue topaz earring accompanied by aquamarines and blue and black sapphires, set in 18-karat white and black gold.

▼ **BLACKENED BLACK GOLD MESH CUFF**

18-karat gold provides contrast for the brilliant white diamonds in this bracelet.

▲ **ARNOLD & SON DSTB**

A true-beat seconds mechanism is exposed on the dial side of Arnold & Son's limited edition DSTB.

AUTHORIZED RETAILER

A. LANGE & SÖHNE
ARNOLD & SON
AUDEMARS PIGUET
BELL & ROSS
BVLGARI
CARTIER
CHOPARD
CLERC
DE BETHUNE
FRANCK MULLER
GIRARD-PERREGAUX
GIULIANO MAZZUOLI
GREUBEL FORSEY
H. MOSER & CIE
HUBLOT
HYT
IWC
JAEGER-LECOULTRE
LUDOVIC BALLOUARD
MAÎTRES DU TEMPS
PARMIGIANI
PIAGET
RICHARD MILLE
ROGER DUBUIS
ULYSSE NARDIN
URWERK
VACHERON CONSTANTIN
WALTHAM
ZENITH

Circular perfection • Focusing on essentials • Harmonious proportions

Patrimony epitomises stylistic purity. Reflecting a deliberately minimalist aesthetic striking a fine balance between taut lines and soft curves, it asserts its personality through slender cases radiating an elegance inspired by the 1950s. Beneath the apparent simplicity, lies a wealth of sophistication.

The fine craftsmanship embodied in Patrimony bears the signature of the prestigious Hallmark of Geneva.

VACHERON CONSTANTIN
GENEVE
SWISS MADE

Letter from the Publisher

Windows on the Sky

Who could have imagined that what began as a way to chart a path through the course of a day would one day harness the wonder of the universe? Nobody—except for watchmakers, of course. Haute horology possesses the miraculous ability to take the unfathomably immense reaches of time and space, and translate this vastness into a diameter of less than 50mm.

Like gateways to the skies above, astronomical complications do not merely illustrate the harmonious dance between Earth and space's celestial objects; they recreate it. With masterpieces contained to dimensions worn elegantly on the wrist, the visionaries of haute horology enable us to observe the dynamic universe around us—a magnificent feat indeed. In the span of a single second, the timepiece's hand travels millions of light-years through space, and through our imagination.

These complications make tangible the mysteries that humanity has unraveled for thousands of years, allowing us to follow and predict celestial patterns, lunar events and our position relative to our neighbors.

This is all actually quite fitting—modern watchmaking was born from a dire need to determine longitude at sea. This challenge required an accurate measure of time against which to reference the position of a known celestial body. Nearly three centuries later, haute horology builds off its momentous past with triumphs of astonishing complexity. Our compulsion to understand the night sky's wonder joins the talent and vision of watchmaking's brilliant virtuosos who allow us to hold the whole universe on our wrist.

Caroline Childers

Caroline Childers

MESURE ET DÉMESURE*
PARMIGIANI FLEURIER
SWISS MADE
TONDAGRAPHE
White gold
Tourbillon chronograph movement
Power reserve indication
Hermès alligator strap
Made in Switzerland
www.parmigiani.ch
PARMIGIANI
FLEURIER
arije
PARIS
50 RUE PIERRE CHARRON +33 (0)1 47 20 72 40 • 30 AVENUE GEORGE V +33 (0)1 49 52 98 88
CANNES
50 BOULEVARD DE LA CROISETTE +33 (0)4 93 68 47 73
SAINT-JEAN-CAP-FERRAT
GRAND-HOTEL DU CAP-FERRAT +33 (0)4 93 76 50 24
LONDON
165 SLOANE STREET +44 (0)20 7752 0246
www.arije.com
* EXACT AND EXULTANT

1.

arije

PARIS
50 RUE PIERRE CHARRON +33 (0)1 47 20 72 40
30 AVENUE GEORGE V +33 (0)1 49 52 98 88
3 RUE DE CASTIGLIONE +33 (0)1 42 60 37 77

CANNES
50 BOULEVARD DE LA CROISETTE +33 (0)4 93 68 47 73

SAINT-JEAN-CAP-FERRAT
GRAND-HOTEL DU CAP-FERRAT +33 (0)4 93 76 50 24

LONDON
165 SLOANE STREET +44 (0)20 7752 0246

Within the everyday world we think we know, we occasionally find pockets of something extraordinary. These places act as a refuge from the outside world, where cares are left at the door and the interior is devoted to a standard of luxury and care unknown anywhere else. Arije is one of those refuges.

With boutique-salons in Cannes, London and Paris, Arije creates worlds within worlds for the watch and jewelry lover who wishes to admire stunning jewels and groundbreaking timepieces within an atmosphere of understated elegance. The refined décor sets the tone of cosmopolitan savoir-faire and hospitality.

The sophisticated members of Arije's multilingual staff are experts in watches and jewelry, as able to discuss carat weight and stunning design as they are the new mechanical triumphs produced each year by such brands as Rolex, Cartier, Vacheron Constantin, Breguet and Guy Ellia. Arije's close relationships with the most prestigious brands in haute horology means that its boutique-salons often offer exclusive limited editions or the very first pieces in highly anticipated collections. Always au courant with the latest developments in horology, Arije has recently added Jacob & Co. and Christophe Claret to its unparalleled selection of watches and jewelry.

3.

1. **OCTO FINISSIMO TOURBILLON**
   —the world's thinnest tourbillon
   (Bulgari)

2. **ROTONDE DE CARTIER ASTROCALENDAIRE**
   (Cartier)

3. **CIRCLE "LA PETITE"**
   (Guy Ellia)

4. **PORTUGUESE SIDÉRALE SCAFUSIA**
   (IWC)

5. **TRI-AXIAL TOURBILLON**
   (Girard-Perregaux)

2.

5.

4.

HARRY WINSTON
CHANEL
Chopard
ZENITH
BVLGARI
CHAUMET PARIS
ROGER DUBUIS
Bell & Ross TIME INSTRUMENTS
BOUCHERON PARIS
FRANCK MULLER GENEVE
Glashütte ORIGINAL
BLANCPAIN MANUFACTURE DE HAUTE HORLOGERIE

BAUME & MERCIER MAISON D'HORLOGERIE GENEVE 1830
Dior
MONTBLANC
JACOB & CO GENÈVE
TAG Heuer
de GRISOGONO GENEVE
TUDOR
OMEGA
JEANRICHARD
MESSIKA PARIS
CHRISTOPHE CLARET
PANERAI

6.

7.

GP GIRARD-PERREGAUX MECHANICS OF TIME SINCE 1791
VACHERON CONSTANTIN Manufacture Horlogère. Genève, depuis 1755.
ROLEX
Breguet Depuis 1775
Cartier
AUDEMARS PIGUET Le Brassus
GUY ELLIA
IWC SCHAFFHAUSEN
PIAGET
HARRY WINSTON
HUBLOT
PARMIGIANI FLEURIER
PIAGET
JAQUET DROZ SWISS WATCHMAKER SINCE 1738
Chopard
AUDEMARS PIGUET Le Brassus
GUY ELLIA
VACHERON CONSTANTIN Manufacture Horlogère. Genève, depuis 1755.
ROGER DUBUIS HORLOGER GENEVOIS
Breguet Depuis 1775
ROLEX
CHRISTOPHE CLARET
ARNOLD & SON SINCE 1764
Cartier
GP GIRARD-PERREGAUX MECHANICS OF TIME SINCE 1791
Glashütte ORIGINAL

8.

JEANRICHARD
IWC SCHAFFHAUSEN
BVLGARI
Bell & Ross TIME INSTRUMENTS
FRANCK MULLER GENEVE
JAQUET DROZ SWISS WATCHMAKER SINCE 1738
CHAUMET PARIS
CHANEL
ZENITH SWISS WATCH MANUFACTURE
de GRISOGONO GENEVE
HUBLOT
PARMIGIANI FLEURIER
Dior
BOUCHERON PARIS
TAG Heuer
BLANCPAIN MANUFACTURE DE HAUTE HORLOGERIE
TUDOR
OMEGA
JACOB & CO GENÈVE
PANERAI
ARNOLD & SON SINCE 1764
MONTBLANC
BAUME & MERCIER MAISON D'HORLOGERIE GENEVE 1830

The flagship boutique-salon of Arije's highly restricted circle, on rue Pierre Charron, has distilled the essence of Arije in Paris's Golden Triangle since 1980. Warm and convivial, it offers more than 30 of the world's most sought-after watch and jewelry brands. Two other stores in Paris on avenue George V and rue de Castiglione (between Place Vendôme and the Jardin des Tuileries in the most elegant area of Paris) add their own perspectives on Arije's vision of luxury. The welcoming ambience of golden light and subtle refinement also pervades the locations in London, Cannes and Saint-Jean-Cap-Ferrat.

6. HOMMAGE DOUBLE FLYING TOURBILLON
(Roger Dubuis)

7. SKELETON TOURBILLON PERPETUAL CALENDAR
(Breguet)

8. PATRIMONY ULTRA-PLATE CALIBRE 1731
(Vacheron Constantin)

9. ROYAL OAK CHRONOGRAPH
(Audemars Piguet)

10. COSMOGRAPH DAYTONA
(Rolex)

9.

10.

Letter from the Editor in Chief

Once you've tried it...

Haute horology has just experienced an exceptional decade. As the rest of the world confronted a severe, tenacious economic crisis that spared only a very few regions and economic sectors across the globe, Swiss watchmaking—and thus haute horology—has made it through with sustained forward momentum. Over 10 years, Swiss watch exports went from 12 billion Swiss francs in 2005 to over 22 billion in 2014, an increase of nearly 100%! This is a mark of the enthusiasm inspired by products that are ever more ingenious, ever more reliable and, above all, ever more understood.

This is the main challenge of haute horology: explain to newcomers this universe, its professions, its expertise and its products. Once the obstacle of ignorance is overcome, nothing can stop the future march of haute horology. Let's not kid ourselves, though: there is still a lot of work to be done. High-end watches are not generally understood; their appeal is not always immediately apparent, while the high price is often seen as a provocation.

These reactions are completely understandable from one who has never visited watch manufacturing facilities, never observed the attention to tiny details necessary for watch assembly procedures, never appreciated how the use of cutting-edge technology works in tandem with traditional hand-craftsmanship. This alchemy is the secret of haute horology, and since the vast majority of customers will not have the chance to visit a manufacturing facility, the media try to capture the magic within, becoming liaisons with the outside world and popularizing a little-known world.

Watchmaking's openness to the wider public and the educational work performed by the watch brands as well as the press have allowed many watch purchasers—once potential, now very much real—to familiarize themselves with this universe and to be persuaded that haute horology timepieces are truly exceptional objects. For more than a decade, Grand Complications has done its part in this project of popularization and explanation. Year after year, we believe that this work continues to be important in garnering new customers. If haute horology is doing better than ever, it is undoubtedly because more people than ever before are informed and educated about what haute horology really is. Because once you've tried it...

Michel Jeannot

Villeret Collection
1735
BLANCPAIN
MANUFACTURE DE HAUTE HORLOGERIE
BLANCPAIN BOUTIQUES ABU DHABI · BEIJING · CANNES · DUBAI · EKATERINBURG · GENEVA · HONG KONG · LAS VEGAS · LONDON
MACAU · MADRID · MANAMA · MOSCOW · MUNICH · NEW YORK · PARIS · SEOUL · SHANGHAI · SINGAPORE · TAIPEI · TOKYO · ZURICH
www.blancpain.com
www.blancpain.tv

GUY ELLIA

PARIS

16 PLACE VENDÔME

« REPETITION MINUTES ZEPHYR »

Premier boîtier au monde en saphir. C'est à travers ce boîtier convexe en saphir taillé dans la masse à très forte résistance et exceptionnelle résonance, que l'on découvre toute la technicité de cette pièce de grande exception. Son mouvement exclusif GEC 88, développé par la manufacture Claret permet de sonner à la demande les heures, les quarts et les minutes sur un timbre cathédral. A ces grandes complications, s'ajoutent l'indication de cinq fuseaux horaires et la réserve de marche.

Worldwide first sapphire case. It's through a sapphire crystal case with an impressive strength and high-resonance that is enhanced the technical nature of this exceptional timepiece. Its' exclusive movement GEC 88, created by the swiss manufacture Claret, chimes on a cathedral tone upon request on the hours, quarters and minutes. Added to these complications, you can read five different time zones and the power reserve.

GRAND COMPLICATIONS®

THE ORIGINAL ANNUAL OF THE WORLD'S WATCH COMPLICATIONS AND MANUFACTURERS®

TOURBILLON INTERNATIONAL
A MODERN LUXURY COMPANY
ADMINISTRATION, ADVERTISING SALES, EDITORIAL, BOOK SALES

11 West 25th Street, 8th Floor • New York, NY 10010
Tel: +1 (212) 627-7732 Fax: +1 (312) 274-8418

Caroline Childers
PUBLISHER

Michel Jeannot
EDITOR IN CHIEF

EDITORS	Elise Nussbaum Julie Singer
ART DIRECTOR	Mutsumi Hyuga
CONTRIBUTING EDITOR	Samson Crouhy
JUNIOR ASSISTANT EDITOR	Amber Ruiz
TRANSLATIONS	Susan Jacquet
VICE PRESIDENT OF OPERATIONS	Sean Bertram
DIRECTOR OF PRODUCTION	Erin Quinn
PRODUCTION COORDINATOR	Tim Maxwell
NATIONAL DISTRIBUTION MANAGER	Hector Galvez
DIRECTOR OF INFORMATION TECHNOLOGY	Scott Brookman
SALES ADMINISTRATOR	Ralph Gago

WEB DISTRIBUTION
www.modernluxury.com/watches

PHOTOGRAPHY
Photographic Archives
Property of Tourbillon International, a Modern Luxury Company

MODERN **LUXURY**

Lew Dickey
CHIEF EXECUTIVE OFFICER

Michael Dickey
PRESIDENT

EXECUTIVE VICE PRESIDENT AND CO-CHIEF OPERATING OFFICER	John Dickey
EXECUTIVE VICE PRESIDENT AND CO-CHIEF OPERATING OFFICER	Jon Pinch
CHIEF FINANCIAL OFFICER	JP Hannan
GENERAL COUNSEL	Richard Denning

TO BREAK THE RULES, YOU MUST FIRST MASTER THEM.
THE VALLÉE DE JOUX. FOR MILLENNIA A HARSH, UNYIELDING ENVIRONMENT; AND SINCE 1875 THE HOME OF AUDEMARS PIGUET, IN THE VILLAGE OF LE BRASSUS. THE EARLY WATCHMAKERS WERE SHAPED HERE, IN AWE OF THE FORCE OF NATURE YET DRIVEN TO MASTER ITS MYSTERIES THROUGH THE COMPLEX MECHANICS OF THEIR CRAFT. STILL TODAY THIS PIONEERING SPIRIT INSPIRES US TO CONSTANTLY CHALLENGE THE CONVENTIONS OF FINE WATCHMAKING.
MILLENARY MINUTE REPEATER
IN PINK GOLD, WHITE ENAMEL DIAL.
AUDEMARS PIGUET
Le Brassus

time

▲ Westime Beverly Hills

Westime

From international timepiece collectors, to customers who are considering purchasing their first watch, Westime's four boutiques in Southern California cater to all by offering a broad and extensive selection of today's most desirable watches.

For nearly three decades, Westime has distinguished itself as the ultimate retail destination specializing in extraordinary watches. John Simonian, a third-generation watch expert with a passion for mechanical timepieces, founded Westime in 1987 when he opened the first boutique on Los Angeles's West Side. From its earliest days, Westime catered to a clientele that ranged from Hollywood celebrities and professional athletes, to the region's influential residents and international business travelers. Westime has since earned the return business of discriminating clients from around the globe.

Today, Westime's four locations reflect the iconic styles of their surrounding neighborhoods. Westime Beverly Hills is an intimate, multi-level boutique at the heart of the city's most glamorous shopping district. The boutique is located on Via Rodeo, a European-style shopping promenade that has emerged as a global destination for the finest watch, jewelry and accessory brands. High ceilings, marble flooring, contemporary furnishings, and showcases that feature champagne water-relief woodwork complement a precisely edited selection of limited editions, high complications and even custom-made timepieces created exclusively for Westime's clientele. Every member of Westime's affable, multi-lingual staff is dedicated to providing exceptional service—from explaining the specifics of complications, to hand-delivering a watch across the country.

Westime La Jolla is located north of San Diego in one of the country's most beautiful seaside communities. The light-filled store resides among charming shops, galleries and cafes on elegant Prospect Street, just steps from the Pacific Ocean. The boutique's gray slate flooring, natural wood and glass watch cases, and open floor plan invite customers to browse casually. The experienced Westime staff provides such special services as watch repairs and water resistance tests.

Westime Sunset is the largest boutique in the family. Its location on Sunset Boulevard in West Hollywood is near the chic restaurants and boutiques of Sunset Plaza and the legendary nightclubs of the Sunset Strip. The 6,600-square-foot Westime Sunset store reflects the bold buildings and signage of the neighborhood. Perforated and backlit metal panels wrap the asymmetrical façade, while a front wall of windows allows passersby to see the brightly lit scene inside. Asymmetrical angles and high-contrast materials including glass, Venetian plaster, steel, walnut and polished concrete create a gallery-like setting inside the two-story space. Custom corners for Audemars Piguet, Breitling, Omega, Zenith, Bulgari, and Buben & Zörweg enhance the shopping experience for fans of those popular brands.

◂ **ROYAL OAK OFFSHORE TITANIUM TOURBILLON CHRONOGRAPH**
(Audemars Piguet)

▲ Westime Malibu

▲ Westime Sunset

Last year, Westime opened its fourth boutique in Southern California, along Malibu's sun-drenched shores. Located within the open-air Malibu Country Mart, Westime Malibu joins such luxury lifestyle neighbors as Lanvin, Ralph Lauren, Calypso St. Barth and Mr. Chow. The boutique's décor complements Malibu's famously laid-back style of luxury. White-washed panels and sea-blue awnings frame Westime's oversize front windows. A cluster of Adirondack chairs provide the perfect spot for shoppers to take in the view of Westime's window displays while enjoying a rest in the sunshine. Inside the boutique, a sand and cream design scheme creates a serene backdrop for back-lit displays, as well as glass cases containing a tightly edited selection of watches and jewelry. Here customers can discover the collections of such prestigious brands as Audemars Piguet, Breitling, Franck Muller, Hublot, Shamballa Jewels and Zenith.

Westime is frequently noted as one of a dozen multi-brand retailers in the world that influences trends in the watchmaking industry. Led by John Simonian, his son Greg and daughter Jennifer, the company is dedicated to offering the most important creations from traditional watch brands, while also promoting the new guard in haute horology. Recent pieces available exclusively at Westime include Audemars Piguet's Royal Oak Offshore Titanium Tourbillon Chronograph limited edition, and Zenith's Pilot Type 20 Annual Calendar Tribute to Russell Westbrook. At Westime's boutiques, there is always something new to discover.

Westime also operates the Richard Mille Boutique Beverly Hills and Hublot Beverly Hills. The company is proud to support numerous charitable causes, including After-School All-Stars, Scripps and Special Olympics.

BRANDS CARRIED

AARON BASHA	H. MOSER & CIE.
AUDEMARS PIGUET	HUBLOT
BELL & ROSS	HYT
BLANCPAIN	JACOB & CO.
BREITLING	LONGINES
BRM	LOUIS MOINET
BUBEN & ZÖRWEG	LUDOVIC BALLOUARD
BULGARI	MAÎTRES DU TEMPS
CHOPARD	MB&F
CHRISTOPHE CLARET	MONTEGRAPPA
DE BETHUNE	NIXON
DEVON	OMEGA
DEWITT	PIERRE KUNZ
DIOR	REUGE
DÖTTLING	RICHARD MILLE
ERNST BENZ	ROLAND ITEN
FIONA KRUGER	RUDIS SYLVA
FRANCK MULLER	SHAMBALLA JEWELS
FREDERIQUE CONSTANT	SNYPER
GIRARD-PERREGAUX	TAG HEUER
GIULIANO MAZZUOLI	TISSOT
GLASHÜTTE ORIGINAL	UGO CALA
GREUBEL FORSEY	ULYSSE NARDIN
HAMILTON	URWERK
HARRY WINSTON	VERTU
HASSELBLAD	VISCONTI
HAUTLENCE	ZENITH

▲ **MAGNUM**
(Buben & Zorweg)

BEVERLY HILLS
254 North Rodeo Drive • Beverly Hills, CA 90210
T: 310-271-0000

LA JOLLA
1227 Prospect Street • La Jolla, CA 92037
T: 858-459-2222

MALIBU
3832 Cross Creek Road • Malibu, CA 90265
T: 310-456-2555

WEST HOLLYWOOD
8569 West Sunset Blvd. • West Hollywood, CA 90069
T: 310-289-0808

info@westime.com / www.westime.com

BIG BANG UNICO MAGIC GOLD
(Hublot)

FIFTY FATHOMS BATHYSCAPHE
(Blancpain)

RM 037 LADIES
(Richard Mille)

Greg Simonian

President of Westime

A WORLD OF WATCHES

WESTIME HAS DOMINATED THE CALIFORNIA WATCH SCENE since 1987, providing a home away from home for the world's most passionate horological enthusiasts. President Greg Simonian is thoroughly versed in watchmaking trends, classic complications and the secrets of matching each client with the perfect timepiece.

Westime has four locations (Rodeo Drive, Sunset Boulevard, Malibu and La Jolla). How do you divide your time among them?

Since my home and our corporate headquarters are also in the Beverly Hills area, I am between our Beverly Hills and Sunset Boulevard boutiques on a nearly daily basis. Malibu and La Jolla are farther afield, but—like our customers—I somehow find a reason to spend entire days working in those boutiques when the beach weather hits its peak!

Do you find the watch-buying culture is different in Los Angeles than in other regions? How so?

Yes, LA-based customers are not afraid to wear oversize watches, and many enjoy pairing sports watches with tailored attire, and classic watches with casual clothing. Those surprising pairings allow the watches to stand out!

How important do you find face-to-face conversations to be when helping a customer make a decision?

With the exception of our clients who have been working with the same Westime salesperson for many years and now enjoy the convenience of shopping with us via phone, email or even text message, the face-to-face factor is very important. And when they see me, the owner, personally standing behind each sale and working to ensure they shop with us for years in the future, it sets Westime apart from less personal points of sale that are part of large corporations.

"When (clients) see me, the owner, personally standing behind each sale and working to ensure they shop with us for years in the future, it sets Westime apart from less personal points of sale that are part of large corporations."

◂ GREG SIMONIAN

What role does the digital world play in Westime's sales? How do you use the Internet and social media to educate potential customers?

At this time Westime does not offer web-only sales—clients can shop from Westime.com and even use our online watch-finder to help narrow their selections. But they eventually need to call or come into our stores to finalize a sale. It has been crucial for Westime to offer useful information on our web site and social media sites too, since customers spend a great deal of time doing their own product research online or on a mobile device today.

To what extent are customers looking for something unusual or even avant-garde, and to what extent are they looking for something with a more traditional aesthetic?

We have many customers who fit both descriptions! That is why Westime places a high priority on offering to customers both the best of traditional horology, as well as collections from independent watchmakers who are finding new and unusual ways of displaying the time.

What exclusive models do you have at Westime? What is the process that goes into releasing an exclusive model?

Among our exclusive models right now are the Audemars Piguet Royal Oak Offshore Titanium Tourbillon Chronograph, a limited edition of 30 pieces, and the Zenith Pilot Type 20 Tribute to Russell Westbrook, a limited edition of five pieces. The process for developing these types of timepieces varies greatly by manufacturer. Of course, for the small independent brands that are only carried at a handful of boutiques around the world—such as MB&F, De Bethune, Fiona Kruger and Ludovic Ballouard—every piece is quite exclusive.

What is the role of site-specific events and dinners, parties, etc, at Westime?

There are many reasons to celebrate at Westime all year! We often host events, ranging from intimate dinners for a couple of clients with a master watchmaker, to watchmaking classes, to cocktail parties in our flagship store to celebrate the US launch of a new watch or the announcement of a celebrity brand ambassador for a line. These events are educational, but they also bring a brand to life for local clients.

What are the trends that you are noticing this year in the watchmaking world? What trends do you see as most exciting for the watch industry?

We will know more after the 2015 watch fairs, but already we are seeing evidence of new materials being used in watches, artistic detailing such as enamel and hand-painting, and the gradual scaling up of smaller watch models.

What new partnerships are you planning for 2015?

At the end of 2014, Westime partnered with Breitling to become the official timekeepers of the Los Angeles Clippers NBA team. So far, it has been a terrific alliance!

Does Westime have any plans for expansion?

We'll have a very active 2015, and that will include the relocation of our Beverly Hills boutique to a new space in the neighborhood.

PIAGET

Perfection in Life

- Jetée des Eaux-Vives, Geneva -
Piaget Emperador Cushion shape 1270P
Ultra-thin automatic tourbillon watch.
Numbered edition.
piaget.fr

Summary

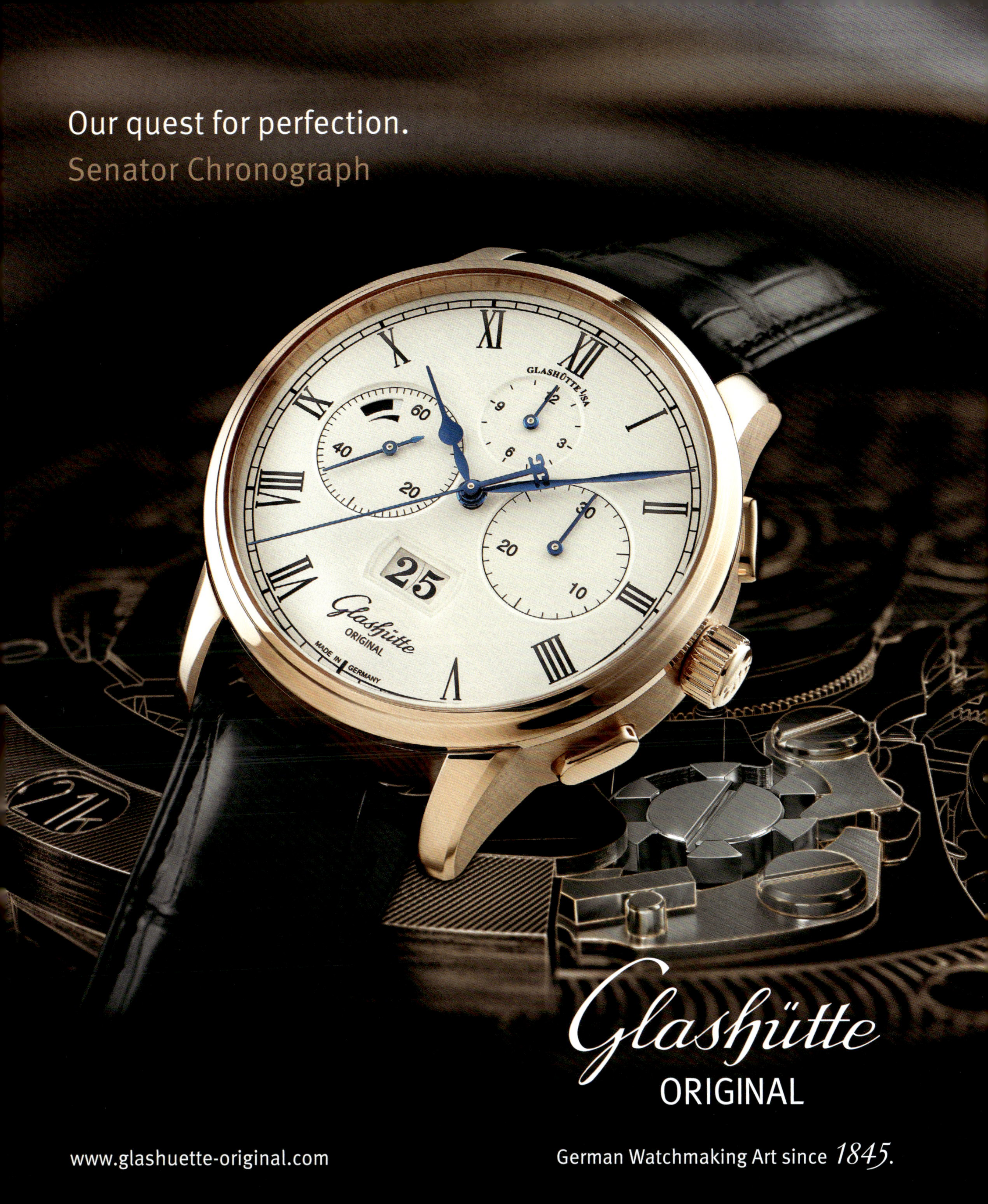
Our quest for perfection.
Senator Chronograph
Glashütte
ORIGINAL
www.glashuette-original.com
German Watchmaking Art since 1845.

the Rolling Stones
EL PRIMERO CHRONOMASTER 1969
TRIBUTE TO THE ROLLING STONES
ZENITH
SWISS WATCH MANUFACTURE
SINCE 1865
WWW.ZENITH-WATCHES.COM | WWW.ROLLINGSTONES.COM

Web Site Directory

A. LANGE & SÖHNE	www.alange-soehne.com
AUDEMARS PIGUET	www.audemarspiguet.com
BELL & ROSS	www.bellross.com
BLANCPAIN	www.blancpain.com
BREGUET	www.breguet.com
CARL F. BUCHERER	www.carl-f-bucherer.com
CARTIER	www.cartier.com
CHANEL	www.chanel.com
CHAUMET	www.chaumet.com
CHRISTOPHE CLARET	www.christopheclaret.com
CORUM	www.corum.ch
DE BETHUNE	www.debethune.ch
DE GRISOGONO	www.degrisogono.com
DIOR HORLOGERIE	www.dior.com
F.P. JOURNE	www.fpjourne.com
FRANCK MULLER	www.franckmuller.com
FREDERIQUE CONSTANT	www.frederique-constant.com
GIRARD-PERREGAUX	www.girard-perregaux.com
GLASHÜTTE ORIGINAL	www.glashuette-original.com
GUY ELLIA	www.guyellia.com
HUBLOT	www.hublot.com
IWC	www.iwc.com
JACOB & CO.	www.jacobandco.com
JAEGER-LECOULTRE	www.jaeger-lecoultre.com
JAQUET DROZ	www.jaquet-droz.com
LONGINES	www.longines.com
PARMIGIANI	www.parmigiani.ch
PATEK PHILIPPE	www.patek.com
PERRELET	www.perrelet.com
PIAGET	www.piaget.com
RICHARD MILLE	www.richardmille.com
ROGER DUBUIS	www.rogerdubuis.com
ROLEX	www.rolex.com
TAG HEUER	www.tagheuer.com
VACHERON CONSTANTIN	www.vacheron-constantin.com
VULCAIN	www.vulcain-watches.ch
ZENITH	www.zenith-watches.com

RELATED SITES

BASELWORLD	www.baselworld.com
SIHH	www.sihh.org

AUCTION HOUSES

CHRISTIE'S	www.christies.com
SOTHEBY'S	www.sothebys.com

J.D
JAQUET DROZ
SWISS WATCHMAKER SINCE 1738
In 1784, the groundbreaking Pierre Jaquet Droz established Geneva's first ever watchmaking factory.
Grande Seconde Quantième Ivory Enamel , ref. J007033200
Ivory Grand Feu enamel, double level. 18-carat red gold case.
Self-winding mechanical movement, silicon balance. Power reserve of 68 hours.
Pointer-type date display at 6 o'clock. Diameter 43 mm.
WWW.JAQUET-DROZ.COM

CHRONOGRAPH
DON'T CRACK UNDER PRESSURE

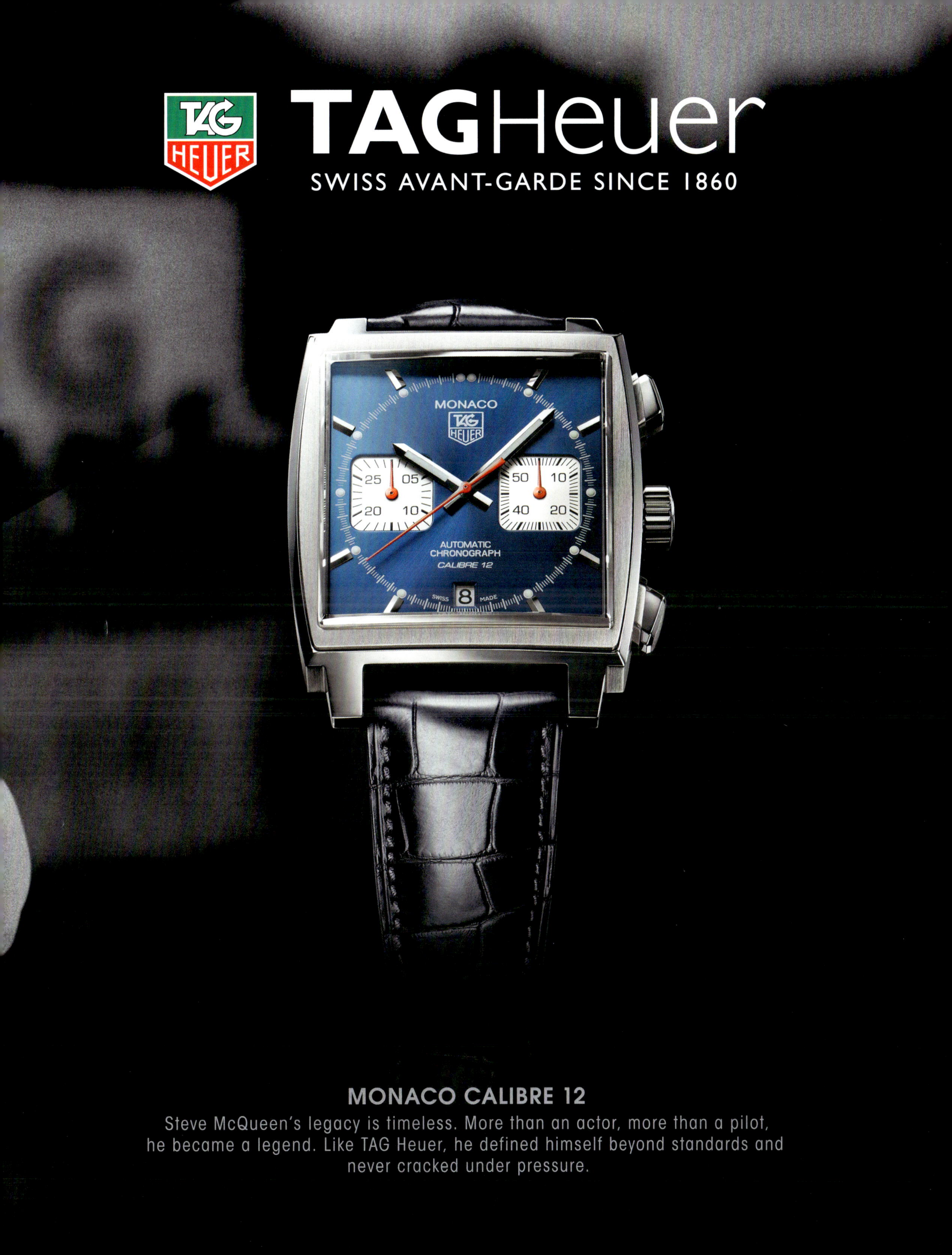
TAG Heuer
SWISS AVANT-GARDE SINCE 1860
MONACO
AUTOMATIC
CHRONOGRAPH
CALIBRE 12
SWISS MADE
MONACO CALIBRE 12
Steve McQueen's legacy is timeless. More than an actor, more than a pilot, he became a legend. Like TAG Heuer, he defined himself beyond standards and never cracked under pressure.

Index

F.P.JOURNE
Invenit et Fecit
Remontoir d'Egalité avec Seconde Morte

JACOB & CO
GENÈVE
NEW YORK • GENÈVE
jacobandco.com

JACOB&CO
NEWYORK 6
NY
L.A.
PARIS
TOKYO
9:12 LA
12:12 NYC
6:12 PARIS
1:12 TOKYO
8:12 MOSCOW

Index

CORUM
LA CHAUX-DE-FONDS · SUISSE
CORUM
ADMIRAL'S CUP
ADMIRAL'S
CUP AC-ONE
ADMIRAL'S CUP REINVENTED, BREAKING THE WAVES FOR MORE THAN 50 YEARS
www.corum.ch

Elegance is an attitude
Simon Baker

LONGINES®
Conquest Classic Moonphase

Index

IWC

JACOB & CO.

JAEGER-LECOULTRE

JAQUET DROZ

LONGINES

PARMIGIANI

VULCAIN
MANUFACTURE DEPUIS 1858
The Watch for Presidents.
NAUTICAL SEVENTIES Limited Edition
Manufacture Vulcain Cricket Calibre V-10.
Hours, minutes, central seconds, alarm,
dial with decompression-table indication.
www.vulcain-watches.com
Manufacture des Montres Vulcain S.A. - Chemin des Tourelles 4 - 2400 Le Locle - Switzerland - Tél. +41 (0)32 930 80 10 - Fax +41 (0)32 930 80 19 - info@vulcain-watches.ch

www.dior.com
Dior
VIII
Dior
DIOR VIII GRAND BAL "PLISSÉ SOLEIL" 36MM.
18 KARAT PINK GOLD, DIAMONDS AND
MOTHER-OF-PEARL. EXCLUSIVE "DIOR INVERSÉ 11 1/2"
AUTOMATIC CALIBRE, 42-HOUR POWER RESERVE.
LIMITED EDITION OF 88 PIECES.

Index

COLLECTIONS

TRADITIONAL COMPLICATIONS WATCHES

Reinterpreting Haute Horlogerie.

For Christophe Claret, the rich heritage of watchmaking is an unlimited field of inquiry. His complications reinvent and revolutionize traditional horology, offering a fresh approach and masterful complexity. Watchmaking is an art form at Manufacture Claret. Each timepiece adds a distinctive, contemporary twist to centuries of traditional watchmaking.

GAMING WATCHES

Transforming time.

Here, time is complemented by original complications turning your wristwatch into a fully-functional casino. These interactive timepieces feature ingenious mechanisms that faithfully replicate the rules of gaming classics. Time is no longer of the essence: it is spent playing at the roulette wheel, at the card table or at games of chance. One thing not left to chance, though, is the quality of the timepieces and their movements.

EXTREME WATCHES

Pushing boundaries of time.

In this collection, cutting-edge research and experimentation govern the design of state of-the-art timepieces in which engineering and mechanics are on full display. Deviating from the beaten track of horology, these avant-garde creations open up new, previously unexplored watchmaking horizons with uncompromising technology.

LADIES' COMPLICATIONS WATCHES

Timespieces that put ladies first.

Carried by elegant and poetic esthetics, stocked with continuously reinvented technical mastery, Haute Horlogerie now comes in a feminine form. Far from being a simple variation of men's watches, with this collection Christophe Claret embodies feminine complexity in these veritable masterpieces.

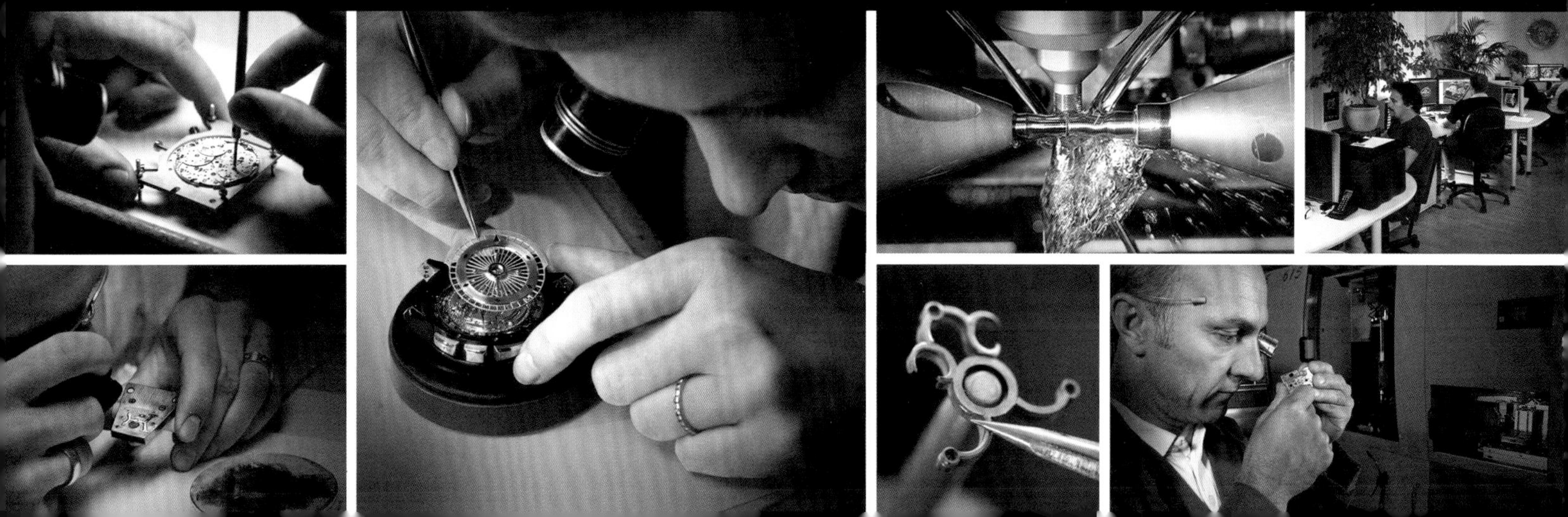

GPHG
GRAND PRIX D'HORLOGERIE DE GENÈVE
2014
Ladies' High-Mech Watch Prize

Margot

FIRST LADIES' HIGH COMPLICATION WATCH, FEATURING «HE LOVES ME, HE LOVES ME NOT...» GAME WITH CHIME AND SYMBOLIC 'COLOUR-FEELING' DESIGNED WINDING ROTOR

RED GOLD CASE
SNOW-SET DIAMONDS

WHITE GOLD CASE
SNOW-SET DIAMONDS

RED GOLD CASE
BAGUETTE-SET DIAMONDS

EXTREME WATCHES

X-TREM-1

WORLD'S FIRST RETROGRADE HOURS AND MINUTES DISPLAY SYSTEM USING MAGNETIC FIELDS AND FLYING TOURBILLON INCLINED AT 30° ANGLE, MOUNTED ON A THREE-DIMENSIONAL MAINPLATE

DUALTOW NIGHTEAGLE

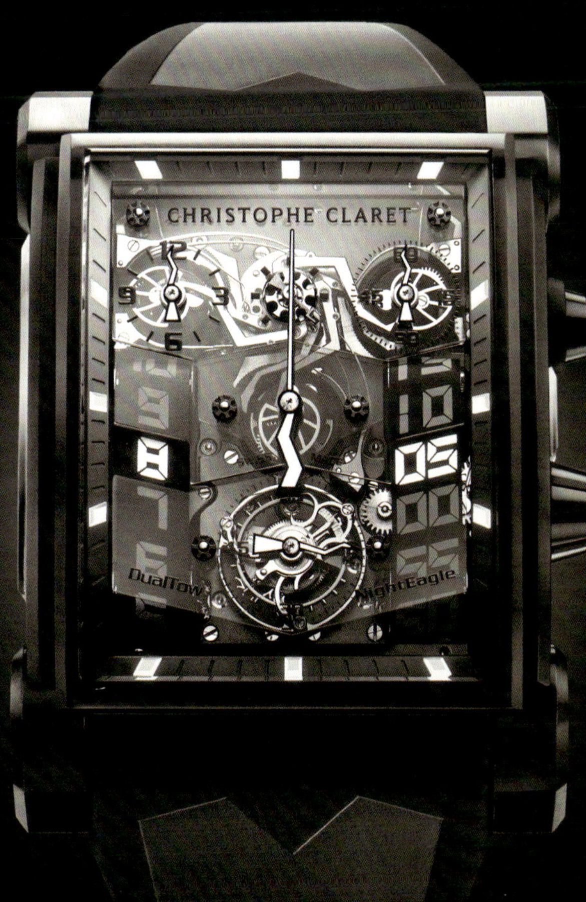

PLANETARY-GEAR, MONOPUSHER CHRONOGRAPH WITH STRIKING MECHANISM, TOURBILLON AND TWO ROLLING BELTS INDICATING HOURS AND MINUTES

2015 NOVELTIES

Allegro

FEATURING MINUTE REPEATER STRIKING THE HOURS, QUARTERS AND MINUTES ON CATHEDRAL GONGS, GRAND DATE AND GMT

AVENTICVM

A TRIBUTE TO THE CAPITAL OF ANCIENT ROMAN HELVETIA, FEATURING A MINIATURE GOLD BUST OF ROMAN EMPEROR MARCUS AURELIUS

Index

Astronomical Watches

Hublot
Antikythera

Astronomical Watches

"The world bemuses me, and I cannot imagine how this 'clock' could exist without a clockmaker," said Voltaire in the 18th century, arguing for the existence of God.

The aptness of this comparison is not merely due to the fact that the organizing principles governing the universe resemble the workings of a clock. It is because horological mechanisms merely reproduce the movements of bodies in space—starting with the Earth and our sun and moon—which form the basis of our system of calendars and other forms of time measurement.

As a sign of this close relationship between horology and the observation of the skies, the first large medieval clocks, as well as the first pocket watches, included several functions related to astronomy and astrology, two disciplines that had not quite become distinct at that point.

With time, the popularity of astronomical indications faded, becoming a specialty reserved almost exclusively for ultra-complicated timepieces.

However, over the last few years, we have been seeing a rekindled interest in this type of complication. In the rush to build upon the rise of the moonphase, a whole host of timepieces have been introduced equipped with celestial maps, miniature planets, indications of the times of sunrise and sunset, equations of time and other functions that sometimes result in quite spectacular dials—especially when it comes to wristwatches.

The reason these astronomical pieces are back in the spotlight is not simply their technical achievements, or because they offer scope for innovation less exploited than that of the tourbillon, perpetual calendar and other traditional complications. It is also because they blend technique and poetry, by reminding us that our conception of time (even as we now measure it, using the periods of radiation of the caesium atom) remains intimately tied to the movement of the heavenly bodies that move ceaselessly above our heads in this vast universe—with or without a divine watchmaker.

Welcome to the universe of outer space, orbiting planets and stardust. Welcome to the world of astronomical watches!

THE SUN WAS OUR FIRST CLOCK. By rising up from the horizon, crossing the expanse of sky, then plunging again into the shadows, it provided us with our first way to divide time: into day and night.

The highest point of its journey marks midday, and the variations in the length of day and night, as well as the return of the seasons, solstices and equinoxes, allowed humans to establish the cycle of the year.

Very quickly, people also noticed that other celestial phenomena also repeated according to a beautifully regular schedule. The lunar cycle, for instance, at roughly 29.5 days, led to our notion of a month. The apparent movement of the stars is another example, with constellations that seemed to move through the night sky and slowly changed position over the year.

As evidence of this attention to the heavens, around 5000 BC large monuments were built whose positioning seems to be closely related to the observation of the movements of the sun and other stars, such as the German "Goseck circle" (4800 BC), or Stonehenge's famous megaliths (3rd millennium BC).

The adoption of writing, around 3400 BC in Mesopotamia, and a century later in Egypt, allowed ancient stargazers to refine their observations and astronomical calculations, to establish increasingly sophisticated calendar systems, to draw the first star maps—and to develop the first horoscopes, using the 12 Zodiac contellations.

Besides sundials and clepsydras, used to measure the passage of time, various instruments designed to observe or reproduce the movement of the planets were developed over the course of millennia and dedicated to the service of religion or of agriculture.

The "Nebra sky disc" (1600 BC), discovered in East Germany, thus depicts the sun, a crescent moon and 32 stars, including—it would seem—the constellation of the Pleiades, a useful tool for determining harvest times.

In antiquity, the Greek scholars of Alexandria devoted themselves to intense research on the sky. One notable such proto-astronomer was Ptolemy, whose "geocentric" system (which placed an immobile Earth at the center of the universe) would prevail until the 18th century. Observation instruments would also undergo great developments, as shown by the famous Antikythera mechanism, the oldest known geared machine (second half of the 2nd century BC). Discovered in 1901 in a shipwreck off the Peloponnesian Islands, this highly complex system facilitated numerous astronomical calculations and calendars linked with the movements of the planets and the Zodiac. Other sophisticated "machines" are described in ancient texts, such as the "moving sphere" attributed to Archimedes that reproduced the movements of the sun, moon and five planets. It is also to the Greeks that we owe the invention of the astrolabe, an instrument allowing astronomers to measure the altitude of celestial bodies and to tell time by reference to the position of the stars or the sun.

In the Middle Ages, Arab scholars built upon ancient knowledge and continued astronomical research, while perfecting very refined mechanisms that were generally powered by hydraulic energy. On the other side of the world, the Chinese also stood out for their development of highly complex machines—such as the famous hydraulic clock created by Su Song (1086), topped by an "armillary sphere" representing the orbits of the sun, the moon and certain stars as seen from Earth.

The birth of mechanical horology in Europe maintained this close relationship with astronomy. The famous Astrarium by Giovanni de Dondi (1364), considered a marvel of the age, displayed on seven dials the movements of the planets known at that time (sun, moon, Mars, Mercury, Jupiter, Venus and Saturn), notably complemented by the sunrise and sunset times.

The large, monumental clocks also gave pride of place to astronomical displays—well before the appearance of minute and second hands. The famous clock on Prague's Old Town City Hall (1410,) for example, possesses a dial with an astrolabe notably displaying the length of day and night, solar time, sidereal time, the signs of the Zodiac and the apparent movements of the sun and moon.

The first portable clocks, as well as the first pocket watches of the Renaissance era, also provided various indications relating to the planets, constellations and the Zodiac—this was a time during which astronomy still went hand in hand with astrology and when any important decision (from a simple business matter to a declaration of war) would be inconceivable without having first "consulted the stars."

The 16th century (with Copernicus), the 17th century (with Galileo and Kepler) and the 18th century would see the solar system pass from a "geocentric" conception to a "heliocentric" model, in which the planets, including the Earth, orbit around the sun.

Starting in the 18th century, timepieces, particularly pocket watches, began to display new astronomical indications of a more scientific character, such as the equation of time, sunrise and sunset times in a given location, star maps, planetary indications or the hours of the tides.

Since then, astronomical displays—less important than previously in everyday life—have effectively become the domain of ultra-complicated watches, used as demonstrations of horological expertise. The Leroy 01 pocket watch (1904) thus displays, in addition to an equation of time, the sky in the Northern and Southern hemispheres represented by three interchangeable discs. The "Graves," created by Patek Philippe in 1933 for the American banker Henry Graves, counts among its functions sidereal time, equation of time and the sunrise and sunset times in New York, as well as a map of the sky visible from Graves's house.

With the renaissance of mechanical horology at the end of the 1980s, watchmakers' attention, as well as that of the public, was first drawn by other complications, such as tourbillons, perpetual calendars or chronographs.

However, over the last few years, astronomical displays—especially the moonphase—have made a triumphant return, both among the big groups and the independent watchmakers such as Christiaan van der Klaauw, who has made the astronomical display a particular focus. Appearing on more and more wristwatches, they allow watch designers to create often quite spectacular pieces that hearken back to our long history of conquering time measurement.

PRINCIPAL ASTRONOMICAL FUNCTIONS

MOONPHASE

A complete description of this function can be found in the "Moonphases" chapter.

SUNRISE AND SUNSET TIMES

The length of day and night varies throughout the year, according to a regular cycle that depends on one's location. This phenomenon is the result of the Earth's tilt on its axis and the shape of its orbit around the sun. On Paris's latitude, the shortest day of the year (winter solstice, December 21) lasts about 8 hours and the longest one (summer solstice, June 21) about 16 hours, while day and night are of equal length (12 hours) on the vernal and autumnal equinoxes (between March 19 and 21 and September 21 and 24, respectively). The closer one gets to the equator, the less variation exists in the length of days and night throughout the year. To reproduce these natural cycles, watchmakers use two cams that complete a complete rotation in one year, with a particular shape befitting the chosen location.

STAR MAPS

Observing the sky continuously over the course of the night, one might notice that the starry skyscape seems to revolve. This apparent movement of the stars is in fact due to the rotation of the Earth. This "rotation" of the stars takes place over a period of 23 hours, 56 minutes and 4.09 seconds (see "Sidereal Time" below), and its axis, as seen from the Northern Hemisphere, seems to be Polaris (also known as the Pole Star or the North Star, from the Ursa Minor constellation), situated almost directly above the axis of the Earth's rotation. However, if one observes the night sky throughout the year at the same time every night, it becomes clear that certain constellations change position over the course of the days and months. This is due to the revolution of the Earth around the sun, and it is why all the constellations of the Zodiac seem to parade by over the year. To reproduce the apparent movement of the stars, watchmakers use a system of discs linked by gears in a setup expertly constructed to be as close as possible to sidereal time. To represent the apparent movement of constellations in the starry night sky over the course of the year, the dial is equipped with a special aperture, whose oval shape outlines, day after day, the portion of the sky visible from a particular spot.

SIDEREAL TIME

If one observes a certain star at midnight as it crosses a particular line of longitude, the following night it will reach the same point at about four minutes to midnight. The time that elapses between two consecutive passages of a star across a line of longitude, or "sidereal time," lasts 23 hours, 56 minutes and 4.09 seconds, corresponding to the time it takes for a complete revolution of the Earth. The difficulty of indicating sidereal time comes from the fact that it does not constitute a round fraction of mean solar time calculated and displayed by the base caliber of the watch. Watchmakers must therefore juggle highly sophisticated gear systems, or even integrate a second, independent system of gears, to achieve maximum precision.

PLANETARIUMS

Moonphases are sometimes completed by the indication of the moon's "age," i.e. the number of days that have passed since the last new moon.

EQUATIONS OF TIME

There are actually two times: true solar time and mean solar time. True solar time, which you might see on a sundial, corresponds to natural rhythms. The Earth describes an ellipse in turning around the sun, and its axis is inclined in relation to its orbit. This means that the lapse of time between two passages of the sun through its highest point in the sky (noon) is not the same length year-round. It actually lasts 24 hours only four times a year: on April 15th, June 14th, September 1st and December 24th. Otherwise, it is sometimes longer and sometimes shorter, following an immutable curve. This difference ranges from 16 minutes and 23 seconds, on November 4th, to 12 minutes and 22 seconds, on February 11th and is called the "equation of time." As for mean solar time (also called civil time), it is a convention based on the average of all days in a year.

Since the variable lengths of true solar days are identically repeated on the same days, they can be "programmed" by means of a so-called "equation" cam that performs one full rotation per year. Its profile resembles that of an "analemma," representing the figure eight traced in the sky by the sun's various positions as recorded each day at the same time and from the same place throughout the calendar year. The extreme precision of this shape determines the accuracy of the equation of time.

The equation of time display may be presented in various ways. Most watches are equipped with a subdial or auxiliary segment swept over by a hand running from -16 to + 14 minutes (sometimes from -15 to +15); to know the true time, the user must then do a mental addition or subtraction of this difference in relation to mean time.

Distinctly more user-friendly yet far more complex in terms of their construction, "running" equation of time watches have two hands fitted coaxially with the minute hand, respectively indicating mean time and true time. They are equipped with a sensor system that searches out the information from the cam, as well as a differential that corrects the running of the supplementary minute hand. They require no mental arithmetic, since true time is continuously displayed directly on the dial.

Star Maps

Mobile and spectacular, star maps provide the wearer with an exact portrait of the night sky in a given location, continually updated and worn on the wrist. Once reserved for ultra-complicated pocket watches, where they would generally occupy the back of the case, these maps are tasking a turn in the spotlight and on the main dial, for wristwatches as well.

Jaeger-LeCoultre, *Rendez-Vous Celestial*

Designed for women's wrists, the Rendez-Vous Celestial from Jaeger-LeCoultre uses its celestial map to display the passage of constellations through the Northern Hemisphere, the signs of the Zodiac and the month on a lapis lazuli dial that bears a deep blue guilloché finishing. A mysterious star-tipped pointer indicates the time chosen for an upcoming rendez-vous.

Patek Philippe, *Star Calibre 2000*

Equipped with 21 complications, the double-sided pocket watch Star Caliber 2000 from Patek Philippe devotes a lot of space to astronomical displays. The front dial is dedicated to the sun and shows a running equation of time, sunrise and sunset times, as well as the daystar's "revolution" over 24 hours. The second dial, on the caseback, depicts the night sky, with a mobile map representing the moon and stars, an integrated moonphase display and a golden ellipse outlining the portion of the sky visible from the place of observation.

IWC, *Portuguese Sidérale Scafusia*

The most complex timepiece ever created by IWC, the double-sided Portuguese Sidérale Scafusia wristwatch displays sidereal hours and minutes onits front dial in addition to solar mean time. The back is given over to a map of the night sky that depicts more than 500 stars and constellations, as well as the horizon, celestial equator and ecliptic (apparent path of the sun on the celestial sphere). This map, which can be personalized according to the geographical coordinates of the client, is complemented by the indications of day, night and twilight, as well as a display of sunrise and sunset times that takes Daylight Saving Time into account.

Vacheron Constantin, *Tour de l'Ile*

Created in 2005 to celebrate the 250th anniversary of Vacheron Constantin, the double-sided wristwatch Tour de l'Ile includes among its 16 complications a display on the caseback dial of a star map, equation of time, sunrise and sunset times as well as a moonphase and moon-age display.

Patek Philippe, *Sky Moon Tourbillon Ref. 6002*

Housed in a white-gold case engraved entirely by hand, the Sky Moon Tourbillon Ref. 6002 double-sided wristwatch from Patek Philippe (12 complications) notably presents on its reverse side a mobile star map representing the apparent movement of the stars and moon (with an integrated moonphase). One can also read the hours and minutes of sidereal time, as well as the meridian passage of Sirius and the moon.

Patek Philippe, *Calibre 89*

The most complicated portable watch in the world (see "Grand Complications" chapter), the Calibre 89 presents several astronomical displays among its 33 complications. These displays are grouped together on the second dial on the back of the watch. An off-center cutout reveals a star map that depicts the Milky Way galaxy and 2,800 stars of the Northern Hemisphere according to their size. A sun-tipped pointer indicates the seasons, equinoxes, solstices and Zodiac signs around the circumference of the dial. The Calibre 89 also displays the hours, minutes and seconds of sidereal time and equation of time, as well as the sunrise and sunset times.

Christiaan van der Klaauw
CVDK Orion

The CVDK Orion from Christiaan van der Klaauw gives pride of place on its dial to a mobile star map rotating in 23 hours, 56 minutes and 4.09 seconds (the sidereal day), with an exact view of the stars and constellations visible from most of the Northern Hemisphere.

Patek Philippe, *Celestial Ref. 6104*

The Celestial Ref. 6104 from Patek Philippe allows the wearer to admire, at any time, the exact configuration of the night sky in the Northern Hemisphere, with the apparent movement of the stars, the position of the moon and its phases over the course of the lunar cycle. The mechanism is composed of three sapphire crystal discs placed one on top of the other, each one 0.2mm thick. A golden ellipse, placed below the crystal, delineates the part of the sky visible above Geneva and all cities on the same latitude.

Jaeger-LeCoultre
Hybris Mechanica à Triptyque

The Hybris Mechanica à Triptyque from Jaeger-LeCoultre (unveiled in 2009 as part of the Hybris Mechanica timepiece trilogy) presents on the back of its reversible case a map of the sky, complete with display of sidereal time, a Zodiac calendar, and equation of time as well as sunrise and sunset times.

Van Cleef & Arpels, *Midnight in Paris*

The Midnight in Paris watch from Van Cleef & Arpels reproduces upon its aventurine dial the movement of the night sky in the heavens above the City of Light. The hinged caseback opens to reveal a calendar set in a small strip of meteorite.

Jaeger-LeCoultre
Master Grande Tradition Grande Complication

The dial of the Master Grande Tradition Grande Complication from Jaeger-LeCoultre displays a celestial map representing the starry sky above the Northern Hemisphere. The small sun-shaped indicator indicates on the circumference of the dial the annual calendar information (date and month) as well as the current Zodiac sign.

Planetariums

Miniature models of the solar system, three-dimensional planetariums allow us to reproduce and admire the movement of the planets, including the Earth, around the sun. Generally constructed as part of larger mechanisms, they have also appeared on a much smaller scale on wristwatch dials.

Richard Mille, *Planetarium Tellurium*

A one-of-a-kind piece that required 10 years of development, the Planetarium Tellurium from Richard Mille—designed with watchmakers Stephen Forsey and Robert Greubel—depicts the revolution of the Earth around the sun (with its rotation on its own axis and its obliquity), the revolution of the moon around the Earth (with rotation on its own axis) and the revolutions of Mercury and Venus. An astrophysicist performed the calculations necessary to reproduce these movements with extreme precision. The considerably enlarged diameter of the model Earth allows it to display distinguishable continents. The moonphases are represented on the moon itself, using a ring to illustrate the part of the moon visible from Earth. The astronomical displays are completed, on the clock face, by an equation of time and a perpetual calendar with many indications (day, date, month, year, decade, leap-year cycle, seasons, equinoxes, solstices, Zodiac signs, etc.).

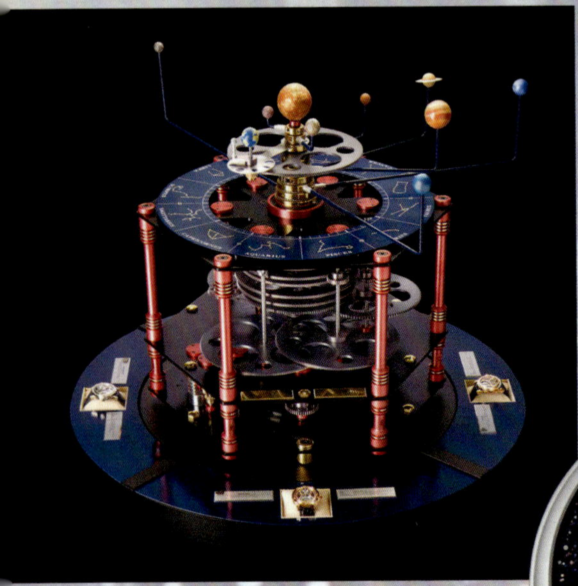

Louis Moinet, *Meteoris Solar System*

The Meteoris Solar System planetarium from Louis Moinet, designed by Rémy Chauvin, depicts an "accelerated" version of the entire solar system, with Mercury, Venus, Earth, the moon, Mars, Jupiter, Saturn, Uranus, Neptune and Pluto orbiting around the sun using about 40 gears powered by a micro-motor. The Earth, for example, revolves around the sun in 37 seconds, as opposed to 365 days. The moon and Mars are covered in dust from, respectively, genuine lunar and Martian meteorites. The annular blue dial indicates the position of each body in the Zodiac sign it traverses. The planetarium is accompanied by four tourbillon watches, each one incorporating a rare meteorite.

Christiaan van der Klaauw
CVDK Planetarium

The Dutch horologer Christiaan van der Klaauw, a renowned specialist in astronomical displays, equipped the CVDK Planetarium with a module that allows it to represent in miniature dimensions the movements of Mercury, Venus, Earth, Mars, Jupiter and Saturn around the sun.

Van Cleef & Arpels, *Midnight Planétarium*

Realized in collaboration with Christiaan van der Klaauw, the Midnight Planétarium from Van Cleef & Arpels depicts on its aventurine dial the movement around the sun of six planets, each one represented by a gemstone, as well as their positions at a given time. The planets move at the speed of their real-life counterparts: Mercury takes 88 days to revolve around the sun, Venus 224 days, Earth just over 365 days, Mars 687 days, Jupiter almost 12 years and Saturn over 29 years. A small shooting star indicates the hours and minutes on a 24-hour scale. The watch's caseback reveals an oscillating mass engraved with a starry décor, as well as two apertures showing the date, month and year.

Christiaan van der Klaauw
CVDK Venus

CVDK Venus from Christiaan van der Klaauw displays on its rotating dial the movement of Venus—the brightest heavenly body after the sun and moon—as well as the revolution of the Earth, around which revolves its own satellite, the moon. The watch also indicates the Zodiac sign in which the sun is positioned.

Graham, *Geo.Graham Tourbillon Orrery*

Equipped with a Christophe Claret movement, the Geo.Graham Tourbillon Orrery from Graham depicts on its face the course of three bodies: Earth (blue sapphire), the moon (rhodium) and Mars (ruby). The position of Earth indicates the date and Zodiac sign.

TRILOGY OF TIME

Ulysse Nardin has revived the astronomical and astrological science of the Renaissance by creating a trio of exceptional timepieces dubbed "Trilogy of Time." The first piece, Astrolabium Galileo Galilei, uses its dial to display the functions of the astrolabe, a navigation tool that led to great discoveries. It indicates the respective positions of the sun, moon and stars, such as we see them from Earth. The Planetarium Copernicus uses a dial composed of seven rings to continuously display the month, Zodiac sign and the astronomical positions relative to the sun and the Earth. As for the Tellurium Johannes Kepler, whose cloisonné enamel dial depicts the Earth as seen from the North Pole, it notably displays the sunrise and sunset times (with dark and light zones) as well as solar and lunar eclipses.

Variations on functions

Distinguishing their creations from watches with star maps or planetariums, some horologers have opted to display only a selected few astronomical functions on their wristwatches—particularly sidereal time and sunrise and sunset times. This approach results in a beautiful blend of technique and poetry.

Van Cleef & Arpels, *Lady Arpels Day and Night*

A beautiful example of a "Poetic Complication™," the Lady Arpels Day and Night from Van Cleef & Arpels possesses a dial that rotates almost imperceptibly over the course of 24 hours, revealing in turn the sun and the moon scintillating on a blue backdrop in aventurine crystal.

Christiaan van der Klaauw, *Real Moon 1980*

In addition to its three-dimensional moonphase at 6 o'clock, the Real Moon 1980 from Christiaan van der Klaauw displays, at 12 o'clock, the height of the sun relative to the horizon using a small mobile sun. At the beginning of summer, the sun is at its highest, and at its lowest at the beginning of winter. The dial at 3 o'clock indicates solar or lunar eclipses.

Christiaan van der Klaauw
CVDK Astrolabium

The CVDK Astrolabium, from Christiaan van der Klaauw, is inspired by the astronomical instrument of the same name, used since antiquity. It provides various kinds of information, including the date, mean solar time, the hours of sunrise and sunset as well as their locations, the position of sun and moon on the ecliptic as well as solar and lunar eclipses.

Hublot, *Antikythera*

A tribute to the famous "Antikythera mechanism," an astronomical calculator dating to the 2nd century BC, the double-sided Antikythera wristwatch from Hublot incorporates the essential indications identified on this ancient masterpiece of engineering. Among them are the Panhellenic Games calendar, the Egyptian calendar, the position of the sun in the constellations of the Zodiac, moonphases, the sidereal year and the display of the different temporal cycles used by ancient scholars.

Arnold & Son, *DBS*

Inspired by two watches created by John Arnold at the end of the 18th century, the DBS from Arnold & Son possesses two small hour-minute dials: one on the left for sidereal time and one on the right for mean solar time, each powered by a separate gear system. The subdial at 12 o'clock serves as a 24-hour indicator for both displays.

Chopard, *L.U.C 150 All-in-One*

Created to commemorate the 150th anniversary of Chopard in 2010, the double-sided L.U.C All-in-One watch combines 14 complications, including a tourbillon and a perpetual calendar. The caseback notably displays an equation of time, a day/night 24-hour indication, the sunrise and sunset times as well as an orbital moonphase—all calibrated to the location of Geneva.

Equations of Time

Extremely rare as a solo complication, the equation of time is most often joined by a perpetual calendar, even an annual calendar, or with astronomical indications. It is usually displayed on a small fan-shaped segment covered by a hand, but certain watchmakers prefer to innovate by spotlighting the complication with a central hand or opting for a linear indication.

Greubel Forsey, *Quantième Perpétuel à Equation*

The Quantième Perpétuel à Equation from Greubel Forsey displays the equation of time on the caseback using an unusual system composed of two transparent discs one on top of the other, which provides a clear visual illustration of the elliptical cycles of the Earth around the sun, with its equinoxes and solstices.

Celestial Fantasies

Inspired by the heavens, a few brands propose fanciful elements that, while not providing rigorous astronomical indications, add a dreamy touch to the dial.

Jacob & Co., *Astronomia Tourbillon*

The Astronomia Tourbillon from Jacob & Co. sets the scene for a tourbillon, an hour/minute dial as well as the Earth and the moon, continuously driven in a circular motion by a four-armed mechanism. The timepiece's spectacular display is not a true astronomical indication, however, as the tiny orbs' movements do not reproduce those of their real-life counterparts.

Louis Moinet, *Astralis*

Besides its "astral tourbillon" and its column-wheel split-second chronograph, the Astralis from Louis Moinet—created in a limited edition of 12 pieces—stands out for its aperture at 6 o'clock that successively displays, over 24 hours, four orbs crafted in extremely rare meteorites: "Dhofar 459" for the moon, "Itqiy" for the sun, "Sahara 99555" for Mercury and "Jiddat al Harasis" for Mars. These pieces of the sky are set in an aventurine disc that effects one rotation per day.

BREGUET

REINE DE NAPLES DAY/NIGHT HIGH JEWELRY – REF. 8999BB-8D-874-DD0D

Combining the arts of horology and jewelry, Breguet's Reine de Naples Day/Night High Jewelry is an impeccable timepiece crafted in an 18-karat white-gold case with a dial in the same material. Diamonds adorn this exquisite piece: 131 baguette diamonds (approx. 5.59 carats) on the bezel and caseband, 73 brilliant-cut diamonds (approx. 0.332 carat) on the flange, one briolette diamond (approx. 0.26 carat) on the crown, a frosting of 35 brilliant-cut diamonds (approx. 0.464 carat) on an attachment and 147 brilliant-cut diamonds (approx. 0.40 carat) on the dial. A complication shows the day and night in the upper half, with the sun represented by the Breguet balance-wheel and a hand-engraved titanium moon, together rotating over the course of 24 hours. The self-winding Cal. 78CS movement provides a power reserve of 57 hours.

HUBLOT

ANTIKYTHERA – REF. 908.OX.1010.GR

Hublot's Antikythera is a technical tribute and a miniature recreation of the entire Antikythera mechanism, an ancient feat of engineering from the second century BC. It also adds a new dimension to this astronomical calculator: that of a timepiece in its own right, capable of accurately measuring the time. The challenge facing Hublot was to integrate a watchmaking movement in a miniaturized reinterpretation of the Antikythera mechanism, respecting the architecture of the original, and in particular its double-sided display. The hours and minutes are displayed in the conventional manner, at the center of the movement recreated by Hublot. This watchmaking movement is regulated by a conventional tourbillon whose cage at 6:00 completes one revolution in one minute. The various known indications of the Antikythera "machine" have been faithfully reproduced on its modern-day counterpart.

JAEGER-LECOULTRE

MASTER GRANDE TRADITION TOURBILLON CELESTE – REF. Q5073401

The Master Grande Tradition Tourbillon Céleste watch, equipped with an orbital flying tourbillon, displays on its dial a full chart of the sky as seen from the Northern Hemisphere. The sun completes one revolution around its rim once every 24 hours, thus indicating not only the time in 24 hours but also the signs of the Zodiac, the days and the months, shown on a circular scale arranged around the edge of the dial. This exceptional spectacle of the watchmaker's art is highlighted by an inner ring set with brilliant-cut diamonds and is adorned by a bezel set with baguette-cut diamonds.

VACHERON CONSTANTIN

PATRIMONY TRADITIONNELLE CALIBRE 2253 – REF. 88172/000P-9495

A hand-stitched dark blue alligator strap with platinum threads traces the path to a truly astonishing spectacle. Limited to 10 numbered pieces, this 44mm 950 platinum timepiece houses a remarkable manually wound caliber honored with the Poinçon de Genève. While the perfectly finished tourbillon with small seconds at 6:00 completes its rotation in precisely one minute, the rest of the dial looks at time on a significantly grander scale. Set to the owner's reference city of choice, the sunrise and sunset indicators surround the exposed carriage on either side. On the upper half of the watch face, the perpetual calendar, complete with day of the week, month, date, and leap-year indicator, is joined by the equation of time between 10:00 and 11:00. Merging the tourbillon's symbolism of horological history with a wealth of highly sophisticated functions, the Patrimony Traditionnelle Calibre 2253 makes a magnificent statement regarding the grandeur of time.

Moonphase

De Bethune
DB25L

Moonphases

Queen of astronomical displays, the moonphase indication has undergone a renaissance over the last few years. Gone are the days of the tiny aperture discreetly submerged among a plethora of subdials and pointers on perpetual calendars or multi-complicated watches. These days, the moon is the main attraction, free to take the stage for a stunning solo on the wrists of men and women alike. To appeal to a wide audience, watchmakers are offering a new take on classics by developing new, original display methods. Chief among these is the trend towards ingenious mobile masking systems or three-dimensional

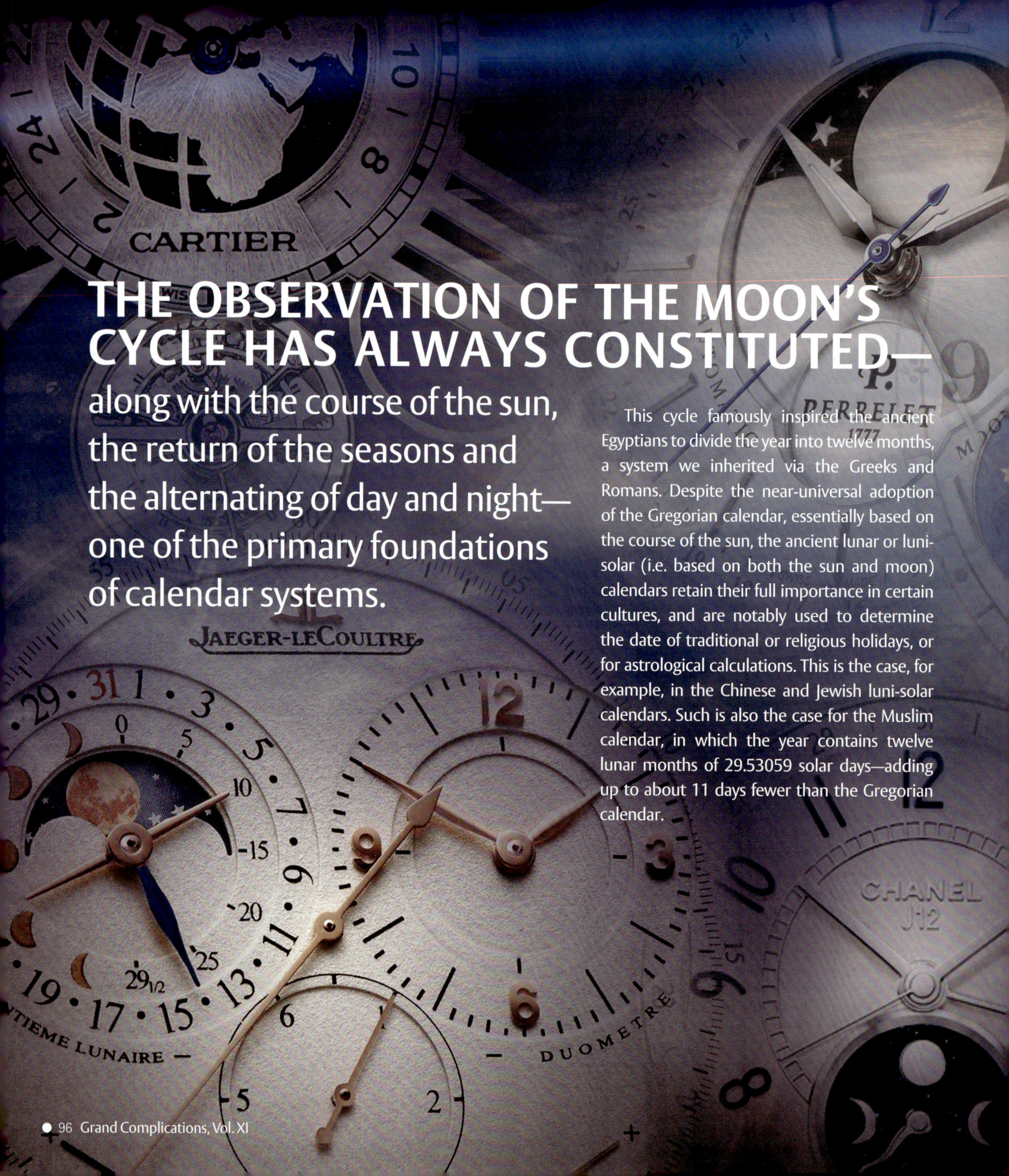

THE OBSERVATION OF THE MOON'S CYCLE HAS ALWAYS CONSTITUTED—

along with the course of the sun, the return of the seasons and the alternating of day and night—one of the primary foundations of calendar systems.

This cycle famously inspired the ancient Egyptians to divide the year into twelve months, a system we inherited via the Greeks and Romans. Despite the near-universal adoption of the Gregorian calendar, essentially based on the course of the sun, the ancient lunar or luni-solar (i.e. based on both the sun and moon) calendars retain their full importance in certain cultures, and are notably used to determine the date of traditional or religious holidays, or for astrological calculations. This is the case, for example, in the Chinese and Jewish luni-solar calendars. Such is also the case for the Muslim calendar, in which the year contains twelve lunar months of 29.53059 solar days—adding up to about 11 days fewer than the Gregorian calendar.

DeWitt, *Twenty-8-Eight Full Moon*

Equipped with a titanium case that houses a partially skeletonized automatic movement, DeWitt's Twenty-8-Eight Full Moon displays the moonphase using a white mother-of-pearl hemisphere that appears and disappears as the days pass, through an aperture reminiscent of an astronomical telescope.

Corum, *Admiral's Cup AC-One 45 Tides*

A watch for water lovers, the Admirals' Cup AC-One 45 Tides from Corum displays—in addition to the moonphase—the time and strength of the tides, as well as the strength of the currents and the height of the tides.

Chopard, *L.U.C Lunar Big Date*

The L.U.C Lunar Big Date from Chopard uses a large aperture at 7:30 to display the moonphase in the Northern and Southern Hemispheres. The "northern" section of the disc depicts Ursa Major, while the "southern" section bears the Southern Cross.

Cartier, *Rotonde de Cartier Jour et Nuit*

The Rotonde de Cartier Jour et Nuit indicates the moonphase using a crescent-tipped retrograde hand. On the top half of the dial, a moon or a sun indicates the hour, completed by a minute hand.

Christiaan van der Klaauw
CVDK Retro Moon

The CVDK Retro Moon from Christiaan van der Klaauw displays all the different moonphases (new moon, waxing crescent, first quarter, waxing gibbous, full moon, waning gibbous, third quarter, waning crescent, new moon again) with a retrograde pointer that sweeps over a sector stretching from 8 to 4 o'clock.

Piaget, *Gouverneur*

Piaget continues its grand tradition of ultra-thin expertise with this Gouverneur model combining a flying tourbillon with an "astronomical" moonphase (showing a one-day discrepancy every 122 years). The small hand precisely follows the waxing and waning phases of the heavenly body.

Arnold & Son, *HM Perpetual Moon*

The HM Perpetual Moon from Arnold & Son bears an oversized moonphase aperture that occupies the entire top half of the dial. The 11.2mm moon stands out for its highly realistic sculpted relief. A second moon on the caseback enables simple and precise adjustment.

Ulysse Nardin
Moonstruck

A worthy heir to Ulysse Nardin's astronomical watches, the Moonstruck continuously indicates the position of the moon and sun relative to any point on the globe; it thus allows its wearer to read the current moonphase and the state of the tides wherever one may be.

Chanel
J12 Moonphase

Made in ultra-resistant high-tech ceramic, the J12 Moon Phase by Chanel features a large aventurine subdial, swept over by a hand progressively pointing to the new moon, the first quarter, the full moon and the last quarter.

Glashütte Original
Senator Panorama Date Moon Phase

Decidedly asymmetrical, the Senator Panorama Date Moon Phase from Glashütte Original displays the moonphase through a large, tilted aperture at 10:30, reflected at 4:30 by the German brand's trademark large Panorama date.

Perrelet, *Moonphase*

The Moonphase by Perrelet gives pride of place to the night star in an aperture occupying the entire lower half of the dial. It is also fitted with a central date pointer.

Round Apertures

Alongside traditional apertures, we are also seeing an increasing number of round apertures, often equipped with mobile covers that represent the dark side of the moon. This enables horologers to create majestic extra-large and even extra-extra-large displays.

Piaget, *Emperador Coussin Moonphase*

On Piaget's Emperador Coussin Moonphase, the disc representing the moon is a plaque of white gold that is heated to create craters such as those found on the face of the moon. Its hand-crafted production renders each watch one of a kind.

Andersen Genève, *Kamar*

For its Kamar model ("kamar" means "moon" in Arabic), Andersen Genève has created a round aperture equipped with a mother-of-pearl disc bearing both a dark and a light moon, all on a dial of aventurine or guilloché 22-karat blue gold.

MB&F, *MoonMachine*

MB&F gave free rein to the Finnish watchmaker Stepan Sarpaneva to create a special version of his HM3 Frog watch. This has resulted in the highly original MoonMachine, with a moonphase display via a round, crown-shaped aperture and a rotor that reproduces the constellations of the Northern Hemisphere.

Bovet, *Récital 9 Tourbillon Miss Alexandra*

On the Récital 9 Tourbillon Miss Alexandra by Bovet, the lunar cycle is displayed by means of a fixed moon with an extremely realistic design that is progressively hidden or revealed by two dedicated discs reproducing the night sky.

Patek Philippe, *World Time Moon Ref. 5575*

Created in 2014 to mark the 175th anniversary of Patek Philippe, the World Time Moon Ref. 5575 is fitted with two mineral glass discs. The lower disc, which depicts the moon, effects one complete rotation over a lunar month. The stationary upper disc is in effect a heart-shaped mask, revealing the visible portion of the moon through a cutout.

Jaquet Droz, *Eclipse Onyx*

On the Eclipse Onyx from Jaquet Droz, a small mobile black disc reproduces the moon's cycle by hiding and revealing a smiling moon, inspired by historical engravings. This spectacle is surrounded by eight applied stars.

Emile Chouriet, *Moonphase*

On the Moonphase model from Genevan watchmaker Emile Chouriet, a large, blue, mother-of-pearl moon moves around the dial, against a backdrop that represents a sky sprinkled with metalized stars.

Cartier, *Rotonde de Cartier Earth and Moon*

The Rotonde de Cartier Earth and Moon is notable for its mechanism that allows the wearer to display the moonphase on command. When the wearer activates the pushbutton at 4 o'clock, a cutout descends and partially obscures the tourbillon carriage. The shape that is thus produced reproduces that of the moon in the sky.

Three-Dimensional Displays

A handful of watchmakers—most often independent artisans—have chosen to supplant the classic aperture systems with innovative three-dimensional displays. In these, the moon is depicted by a small mobile sphere (or a stationary sphere equipped with a mobile cover) reproducing the different faces of the moon throughout its cycle. This leads to very evocative stagings.

Konstantin Chaykin, *Lunokhod*

On the Lunokhod from Russian watchmaker Konstantin Chaykin, the heavenly body is represented by a stationary sphere sculpted in wootz steel, an Indian metal whose granular quality recalls the lunar surface. The moonphase is displayed by means of a semi-circular cover in black rhodiumed silver, which rotates around a small globe..

VicenTerra, *Luna*

The VicenTerra Luna is adorned with two small rotating globes. The one on the left shows the moonphase, while the one on the right indicates at what location on the Earth it is currently noon. The very top of the dial depicts a sunrise and a sunset, followed by a starry night.

Cyrus, *Klepcys*

The Klepcys, designed by Cyrus in collaboration with the watchmaker Jean-François Mojon, boasts a black disc that slowly rises to hide a spherical moon, dotted with craters. During the new moon, the orb gives way to the brand's logo.

De Bethune, *DB28 Skybridge*

The DB28 Skybridge by De Bethune combines a patented spherical moonphase with a night-sky dial in blued and mirror-polished grade-5 titanium, studded with white-gold and diamond stars.

Thomas Prescher

Mysterious Automatic Double Axis Tourbillon

On the Mysterious Automatic Double Axis Tourbillon from Thomas Prescher (whose movement is entirely hidden within the bezel), the two cylinders for the hours and minutes frame a spherical moonphase display.

The Feminine Side of the Moon

The moonphase display remains one of the favorite complications of ladies' watches, though it is still more widespread on masculine wrists. Catching the wave of this longstanding interest, both large brands and independent watchmakers offer a beautiful palette of quite refined ladies' models, most often adorned with diamonds and mother-of-pearl.

Patek Philippe
Diamond Ribbon Ref. 4968

Illuminated by a diamond spiral that wraps around the case, Patek Philippe's Diamond Ribbon Ref. 4968 features a highly precise moonphase (accumulating a one-day discrepancy every 122 years) upon its mother-of-pearl dial.

Breguet, *Reine de Naples*

Inspired by the wristwatch created for Napoleon's sister Caroline Murat, the Reine de Naples from Breguet has been introduced in a self-winding model with moonphase and a power reserve display.

Zenith, *Star Moonphase*

The Star Moonphase from Zenith—equipped with the famous, ultra-thin, automatic Elite movement—features an imposing moonphase at 6 o'clock upon a dial punctuated by large blue Roman numerals. The cushion-shaped case is available in steel or rose gold.

Corum, *Admiral's Cup Legend 38 Mystery Moon*

A magical ballet: as it spins steadily around, the finely worked mother-of-pearl dial upon the Admiral's Cup Legend 38 Mystery Moon from Corum moves the date aperture (which features the sun) as well as the moonphase across the dial.

Saskia maaike Bouvier, *The 8 Moons*

Created by a woman for women, The 8 Moons timepiece from Saskia maaike Bouvier sports a starry night sky as a backdrop for eight spheres that show not only the current phase of the moon (at 12 o'clock), but also those for the seven following nights.

Graff Diamonds, *Diamond GyroGraff*

Symbiosis of horology and haute joaillerie, the Diamond GyroGraff from Graff Diamonds features a three-dimensional moonphase indicator and a bi-axial tourbillon on a dial encircled by a bezel with a patented invisible setting of diamonds.

Records for Precision

On ordinary moonphase watches, the complication will eventually show a one-day discrepancy after two years, seven months and around 20 days. On the moonphase complications within the luxury watches at the high end of the market, that difference has been reduced to one day over 122 years and 44 days. A few great brands have opted to face this technical challenge and develop moonphases that are much more precise—even resulting in no lag at all.

IWC, *Portuguese Perpetual Calendar*

Using an innovative system comprised of thee toothed wheels, the moonphase indication on IWC's Portuguese Perpetual Calendar has a discrepancy of just one day over 577 years. The double hemisphere display for the Northern and Southern Hemispheres is completed by two countdowns signaling the number of days until the next full moon.

H. Moser & Cie, *Moser Perpetual Moon*

The Perpetual Moon from H. Moser & Cie would need to run 1,027.3 years to differ from the real lunar cycle by one full day—thanks to a continually revolving display linked to the hours mechanism (as opposed to most moonphases, which advance once per day). The vertical lines marking the eight quarters of the moon, fixed by international standards, allow an ultra-precise reading.

A. Lange & Söhne, *1815 Moonphase*

The German company A. Lange & Söhne has equipped its 1815 Moonphase with a gear train fitted with a particular gear-train ratio that guarantees a difference of just 6.61 seconds per lunar month, or just one day every 1,058 years.

ochs und junior, *Moon Phase* 3478

27 years: with the Moon Phase model from the youthful ochs und junior brand, the famous watchmaker Ludwig Oechslin has created an extremely precise moonphase indication while using just five components to modify the base movement. The dial also indicates the position of the moon in relation to the Earth and the sun.

Andreas Strehler, *Sauterelle à Lune perpétuelle*

On its Sauterelle à Lune perpétuelle, Andreas Strehler has designed a moonphase indication that needs to be corrected by one single day after 14,189.5383 years.

Christiaan van der Klaauw, *CVDK Real Moon Joure*

The CVDK Real Moon Joure from Christiaan van der Klaauw—specialist in astronomical watches—features a three-dimensional, handmade moon at 6 o'clock, which deviates from the lunar cycle by just one day every 11,000 years.

A. LANGE & SÖHNE

GRAND LANGE 1 MOON PHASE – REF. 139.032

Positioned prominently within the watch's principal subdial at 9:00, the hand-wound L095.3 caliber's moonphase takes on the lunar cycle with bold ingenuity and distinguished sophistication. The intricately coated night sky, adorned with more than 300 stars of varying sizes, depicts a vivid moon that, thanks to its connection to the caliber's hour-wheel continuum, is in constant motion through its monthly journey. In addition to this touch of undetectable, yet extraordinary, mechanical detail, the timepiece is endowed with a seven-gear transmission that reduces the lunar cycle's deviation to less than one minute, requiring a mere one-day correction every 122.6 years. The 41mm pink-gold wristwatch is finished with a large date, subsidiary seconds and a 3:00 indicator of the movement's 72-hour power reserve.

CHANEL

J12 MOONPHASE – REF. H3407

A watchmaking icon of the 21st century, the J12 watch has reinvented itself season after season while still respecting the essence of its design. A heavenly body illuminated by the sun and a symbol of femininity, the moon has been veiled in mystery since the dawn of time. The elegant silhouette in high-tech ceramic reveals, in an aventurine starry sky, the phases of the moon that reveal their secrets to those who know how to observe them.

The J12 Moonphase is also available in white and black high-tech ceramic classic or set with diamonds.

DE BETHUNE

DARK SHADOWS – REF. DB28ZC8

De Bethune now offers a matte black interpretation of its iconic DB28 model, available in a limited edition of 50 pieces. The DB28 Dark Shadows opts for a matte, satin-brushed zirconium case. Its dial opens a window onto the movement, displaying a wealth of nuances and revealing a variety of shades ranging from anthracite gray to intense black. The DB28 is equipped with Calibre DB2115, endowed with De Bethune's numerous technological innovations. Its precision is further consolidated by the self-regulating twin barrel, silicon/palladium balance wheel and balance-spring with a flat terminal curve, protected by the triple pare-chute system. The spinning moonphase display uses a palladium and anthracite zirconium hand-polished steel sphere boasting a degree of precision equivalent to a mere one-day difference every 122 years.

DE BETHUNE

DB25L JEWELLERY – REF. DB25SJWS3

This creation is distinguished by the many ways in which the light plays across the materials and the complex dial architecture. The dark gleam of hand-polished and blued titanium, the sparkling twinkle of the white gold and diamond stars, and the muted glow of the curved blued steel hands make for a stunning piece. The depth of this star-studded sky, which can be personalized with date and location is revealed through the alternating curves that are convex on the outside for the chapter ring, and concave in the center. A spherical moon set with sapphires and diamonds shines at 12:00, providing an exceptionally accurate moonphase display that will diverge from astronomical reality by just one day every 122 years.

DE BETHUNE

DB25L – REF. DB25LWS3V2

Powered by a manual-winding DB 2105 V2 caliber, this model is equipped with 27 jewels and a power reserve of 6 days, and beats at 28,800 vph. This exceptional timepiece is housed in a shining white-gold drum-shaped case with cone-shaped lugs and a crown at 3:00. The bold blue hue of the De Bethune star-studded sky on the dial, which can be personalized, sets the perfect display for the platinum and flame-blued steel spherical moonphase indicator at 12:00. This watch is finished with an extra-supple alligator leather strap with a pin buckle.

DE BETHUNE

DB28 SKYBRIDGE – REF. DB28CE

Ancestral skills and the latest scientific breakthroughs merge to make this timepiece a perfect summary of highly aesthetic ambition and peerless technical precision. While classically inspired in terms of its round shape, its crown at 12:00 and its hunter-type back borrowed from pocket watches, the ultra-light mirror-polished titanium case of the DB28 with its distinctive floating lugs nonetheless proclaims the collection's modern and indeed futuristic nature. An eminent reference to 18th-century clocks, the star-studded sky in mirror-polished and blued titanium derives its power and beauty from its concave shape and multiple decorations. The spherical moonphase display is accentuated by an arrow-shaped bridge that appears to be pointing towards infinity.

DIOR

CHIFFRE ROUGE C03 – REF. CD084C11A003

The Chiffre Rouge collection is enriched by a new moonphase watch, equipped with an Elite movement made by Zenith manufacture. Reinterpreting facets of the Dior Homme universe and its codes, the moonphase indicator at 6:00 and the small second counter at 9:00 combine shades of black, gray and pink gold while the large date is displayed red at 2:00. This elegant timepiece, with a gray metallized mother-of-pearl dial, is available in a limited edition of 100 pieces.

F.P. JOURNE

OCTA LUNE

The Octa Lune with automatic winding offers over 120 hours of chronometric power reserve, and reminds us of ancient times when the full moon was the only source of light during the night. While today we no longer depend on moonlight for night vision, the moonphase indication remains one of the most poetic of all horological complications. The hours and minutes are placed right of center for a quick and discreet reading, with small seconds, large date, power reserve indication and moonphase rounding out the dial. The platinum case joins a white-gold and silver dial, and movement in 18K rose gold.

GLASHÜTTE ORIGINAL

PANOMATIC LUNA – REF. 90-12-01-12-04

This ultra-feminine timepiece has a beautiful light-colored mother-of-pearl dial set with 18 diamonds. The stainless steel case is also set with diamonds, framing the delicately placed functions on the dial. The hour and minute counter is off-centered, as is the small seconds counter. There is also a moonphase at 2:00 and a signature Glashütte Original Panorama Date at 4:00. This piece rests on a white rubber strap and is powered by a Caliber 90-12 with a power reserve of 42 hours.

GLASHÜTTE ORIGINAL

PANOMATICLUNAR – REF. 90-02-42-32-05

The Pano collection's signature asymmetrical effect is easily appreciated with a glance at the new PanoMaticLunar, which features the characteristic off-center visuals. The hour/minute and small seconds subdials are aligned along a vertical axis to the left, while the Panorama Date and superb moonphase display are positioned to the lower and upper right, respectively. The stainless steel case frames a warm silver dial with silver hour indexes beneath blued steel hands, complemented by the moonphase display and characteristic Panorama Date. Within beats the manufacture's superb Calibre 90-02, a self-winding 47-jewel movement with a power reserve of 42 hours. The sapphire crystal caseback provides a clear view of the finely finished movement, including the typical Glashütte three-quarter plate with Glashütte ribbing, off-center skeletonized 21K gold rotor, blued screws and hand-engraved balance cock.

IWC

PORTOFINO MIDSIZE AUTOMATIC MOON PHASE – REF. IW459002

Much like the Roman numerals, feuille hands and convex sapphire crystal, the moonphase display is an integral part of the Portofino watch line's identity. The celestial bodies are not the only stars on the dial—the Portofino Midsize Automatic Moon Phase features an exquisite collection of diamonds and the finest materials. With a moderately sized 37mm 18K red-gold case, the watch has ideal dimensions for a more slender wrist. The hands and moonphase display are powered by the 35800-calibre automatic movement, and the moonphase can be corrected using a button at 2:00. IWC's watch designers drew inspiration from midcentury Hollywood glamour for the flawless, gently shimmering white mother-of-pearl on the dial. A bezel set with 66 diamonds completes each piece with a fitting elegance.

PATEK PHILIPPE

LADIES COMPLICATIONS – REF. 4968G-010

Created in a cool color scheme, with a shiny taupe alligator leather strap and white mother-of-pearl dial, this Ladies Complications timepiece from Patek Philippe is set with 273 graduated diamonds in a spiral (approximately 2.12 carats) on its 33.3mm white-gold case. Continuing the spiral motif, the dial bears a spiral engraving that swoops around the gold ruthenium-black numeral appliqués. The moonphase at 6:00 also includes a small seconds display. This beautiful piece is powered by the manual-winding 215 PS LU caliber, awarded with the Patek Philippe Hallmark, whose 157 components and 18 jewels fit within a diameter of 21.9mm and a thickness of 3mm.

PATEK PHILIPPE

LADIES COMPLICATIONS – REF. 7121/1J-001

Worn on a yellow-gold bracelet, this seductive ladies' timepiece shimmers and reflects light in every direction. Housed in a 33mm 18K yellow-gold case illuminated by the shimmer of 66 flawless Top Wesselton diamonds, the ultra-thin hand-wound 215 PS LU caliber boasts a highly precise moonphase complication that needs only one adjustment every 122 years. Combined with a hand-guided small seconds display for an optimal use of space, the golden moon is exhibited upon a starry sky in a subdial at 6:00. The hours and minutes are indicated via two "Poire Stuart" hands in 18K gold, against 11 Breguet numerals in the same precious metal, all on a grained silvery white dial. This model is endowed with the Patek Philippe Hallmark.

ZENITH

EL PRIMERO CHRONOMASTER GRANDE DATE – REF. 03.2160.4047/01.C713

The self-winding El Primero 4047 caliber, revealed through a large aperture in the upper-left quadrant of the silver-toned sunray dial, drives a poetic display of both the moon and sun phases. The exquisitely decorated opening, at 6:00, animates the double path of the two celestial objects by way of two superimposed discs. As the upper disc performs a full rotation every 59 days, the lower disc, representing the day/night status, completes its revolution once per day. At 2:00, two concentric discs are used to display the date in a large and legible format. A 30-minute counter, at 3:00 on the silver dial, complements a column-wheel chronograph whose seconds are indicated to the tenth of a second by a slender central blue stem.

Minute Repeaters

Vacheron Constantin
Maîtres Cabinotiers Skeleton
Minute Repeater

Minute Repeaters and

Seiko, *Credor Spring Drive Minute Repeater*

Signed Seiko, the Credor Spring Drive Minute Repeater features a skeletonized movement that is visible through an openworked dial and a sapphire crystal caseback. Its hammers are shaped from a special kind of steel forged by one of the most famous Japanese steelmakers.

Bovet
Amadeo Fleurier 46 Minute Repeater Tourbillon Reversed Hand-Fitting

Equipped with a visible movement that is entirely engraved, the Amadeo Fleurier 46 Minute Repeater Tourbillon Reversed Hand-Fitting from Bovet can easily be transformed into a table watch, pocket watch or pendant watch.

Repeater Watches: New Rhythms

The most popular repeater watches are currently the minute repeaters, which sound on demand the hours, quarter-hours and minutes. A few horologers have opted to reinterpret this classic model by revisiting the old-fashioned "five-minute repeater" (simpler to produce and hence less costly), proposing variations on the frequency and type of the strike, or creating repeaters that sound the time of another time zone.

Claude Meylan, *Répétition 5 (Ref. 8731)*
The Répétition 5 (Ref. 8731) by Claude Meylan features a movement entirely openworked and engraved by hand. Pressing the pushbutton at 8 o'clock makes the watch sound the number of hours, followed by the number of five-minute intervals elapsed since the previous hour.

Patek Philippe, *Chiming Jump Hour Ref. 5275*
Created in honor of Patek Philippe's 175th anniversary, the limited edition Chiming Jump Hour Ref. 5275 features three instantaneously jumping indications (digital hour, pointer-type minutes and seconds) and a striking mechanism that discreetly sounds each full hour with a single chime.

Seiko, *Credor Spring Drive Sonnerie*
The Japanese brand Seiko has revived the antique hour repeater mechanism in a high-end model named Credo Spring Drive Sonnerie. The mechanism strikes in three modes: the number of hours on each hour; three strikes every three hours (3, 6, 9 and 12 o'clock); and silent mode.

Louis Vuitton, *Tambour Répétition Minutes*

On the Tambour Répétition Minutes from Louis Vuitton, with a dual time zone, the minute repeater sounds the home time, not local time—a pleasant way to feel connected with close friends and family.

Kari Voutilainen, *Masterpiece N° 7*

On his one-of-a-kind model, the Finnish-born watchmaker offers a minute repeater striking not only hours/quarter-hours/minutes, but also hours/ten minutes/minutes, a system designed to enable a more intuitive division of time.

deLaCour, *Birépétition minutes*

Equipped with a Christophe Claret movement, the Birépétition minutes from deLaCour strikes the hours, quarter-hours and minutes for either the local time or a second time zone.

Grand and Small Strikes

Like a church bell on the wrist, grand and small strike systems, also called grande et petite sonnerie, mark the hours and quarter-hours in passing (with grand strikes repeating the hours each time). This extremely rare function is the prerogative of ultra-complicated wristwatches (see chapter on Multi-complications). Having first appeared in wristwatch form in a model presented by Philippe Dufour in 1992, it has also cropped up since then in some highly sophisticated watches where it most often joins the minute repeater—sometimes in chiming the famous melody of Big Ben.

A. Lange & Söhne
Zeitwerk Striking Time
The Zeitwerk Striking Time from A. Lange & Söhne strikes each quarter-hour in passing with a high note, then the hour with a low one—thus reproducing in slow motion the famous tune from the beginning of Beethoven's Fifth Symphony.

Patek Philippe, *Grandmaster Chime*
Designed for the 175th anniversary of Patek Philippe (1839-2014), the Grandmaster Chime wristwatch presents, among its 20 complications, no fewer than five acoustic functions, including a grand strike, small strike, minute repeater, on-demand date repeater and an alarm that strikes the alarm time.

Bulgari, *Grande Sonnerie Perpetual Calendar*
The Bulgari Grande Sonnerie Perpetual Calendar has a chime playing the Westminister tune on four notes. It is also equipped with two barrels, of which one is exclusively dedicated to the striking mechanism and guarantees a 24-hour power reserve in grande sonnerie mode.

Philippe Dufour, *Grande Sonnerie*

Launched in 1992, the Grande Sonnerie from Philippe Dufour—one of the preeminent members of the Académie Horlogère des Créateurs Indépendants—was the first minute repeater watch with grand and small strike.

Audemars Piguet, *Grande Sonnerie Carillon*

Presented at the end of the 1990s, the Grande Sonnerie Carillon by Audemars Piguet is distinguished by its dynamograph system indicating the real torque delivered by the barrel. It has a power reserve of 20 hours in Grande Sonnerie mode and 60 hours in Petite Sonnerie mode.

F. P. Journe, *Sonnerie Souveraine*

Possessing grand strike and minute repeater capabilities, the Sonnerie Souveraine is distinguished by its extreme functionality and security assured by ten patented devices. F.P. Journe actually designed it "so an eight-year-old child could use it without damaging it"—which required a high degree of technical sophistication in the conception and design.

Watches with Automatons

In terms of a grand visual and acoustic spectacle, watches with automatons, which are extremely rare, unquestionably occupy an echelon all to themselves.

Van Cleef & Arpels, *Lady Arpels Poetic Wish*

Signed Van Cleef & Arpels, the Lady Arpels Poetic Wish combines a five-minute repeater with three automatons—a young girl, a tiny mother-of-pearl cloud and a kite—moving on request to sound the hours and minutes in music.

Jaquet Droz, *The Bird Repeater*

Heir to the famous 18th-century Jaquet Droz automatons, The Bird Repeater wristwatch features a pair of birds on a nest in front of a stream—a three-dimensional tableau that springs to life each time the cathedral-gong minute repeater is activated.

Ulysse Nardin
Jazz Minute Repeater

The Jazz Minute Repeater by Ulysse Nardin associates traditional horology with the art of automaton jaquemarts (jacks), by means of four small musicians performing an animated concert each time the wearer triggers the minute repeater.

Bovet
Amadeo Fleurier 44
Minute Repeater Tourbillon
Triple Time Zone Automaton

The Amadeo Fleurier 44 Minute Repeater Tourbillon Triple Time Zone Automaton watch by Bovet features a dial featuring an automaton with bells reminiscent of the first jaquemart (jack) mechanisms.

Bulgari, *Commedia dell'Arte*

The Commedia dell'Arte by Bulgari sets the stage for the three main protagonists in this traditional form of Italian theatre. When the "cathedral gong" minute repeater is activated, the central figure (whose arm shows the minutes) raises his right arm, while the other figures begin a lively dance.

Bulgari, *Daniel Roth collection, Il Giocatore Veneziano*

Inspired by a painting by Caravaggio, Il Giocatore Veneziano from Bulgari's Daniel Roth collection associates a minute repeater mechanism (with cathedral chime) with an automaton that throws two dice with a total of 504 possibilities. The automaton movement may be activated either at the same time as the striking mechanism, or on demand.

Feminine Melodies

In a still-tentative trend, we have recently seen the first watches with minute repeater or "in passing" strike mechanisms created especially for women.

Patek Philippe
Ladies First Minute Repeater
Ref. 7000

The first Patek Philippe minute repeater wristwatch exclusively dedicated to women, the Ladies First Minute Repeater Ref. 7000 is distinguished by its ultra-thin self-winding movement and its pure aesthetic, featuring a rose-gold case set with diamonds.

Ellicott, *Lady Tuxedo Midnight*

Ellicott's Lady Tuxedo Midnight model is equipped with a self-winding movement striking the hours and quarter-hours on request. Its sensual case, partially or entirely set with diamonds, is crafted in rose gold, white gold or steel.

Breguet, *Reine de Naples Sonnerie au passage 8978*

To celebrate the 200th anniversary of the wristwatch created for Caroline Murat, Queen of Naples, Breguet presented the Reine de Naples Sonnerie au Passage 8978, with a mechanism that marks each full hour with two strikes repeated three times.

Musical Watches

To ring out the melody of time, several watches incorporate miniature mechanisms inspired by historical music boxes. They feature a varied repertoire ranging from great classic tunes to modern hit songs.

Charles Reuge
Reuge CXH.17.4101.000

The renowned Swiss music box specialized manufacturer Reuge has equipped the hunter-type Charles Reuge pocket watch with a miniaturized mechanism playing an extract from Mozart's The Magic Flute or A Little Night Music, while the automatons spring to life around the fountain.

Boegli, *M.63*

Boegli has equipped its M.63 Far West watch with a "mechanical music movement" playing the famous tune "For a Few Dollars More" by Ennio Morricone. The brand also offers various wristwatches and pocket watches paying tribute to the great composers such as Verdi, Mozart and Vivaldi—even "God Save the Queen."

Ulysse Nardin, *Stranger*

On its mechanism, clearly visible through the dial side, the Stranger watch by Ulysse Nardin plays the tune of "Strangers in the Night" every hour or whenever the user presses a button. The crown with its integrated pusher serves to adjust various functions, including winding, moving the date forward or back, as well as setting the time.

A. LANGE & SÖHNE

LANGE ZEITWERK STRIKING TIME – REF. 145.029

This member of the Lange Zeitwerk family is endowed with a chiming mechanism that is visible on its dial. The Lange Zeitwerk Striking Time is the first Lange wristwatch with an acoustic signature. It strikes the quarter-hours with high-pitched tones and the full hours at a lower pitch. The chiming mechanism consists of two black polished steel hammers that are integrated in the dial layout on either side of the subsidiary seconds. The manually wound L043.2 caliber is crafted to the most exacting Lange quality standards and largely decorated and assembled by hand. The sonorous timepiece comes in a 44.2mm white-gold case with a black dial or in a limited edition of 100 pieces with platinum cases and rhodium-plated dials.

A. LANGE & SÖHNE

LANGE ZEITWERK STRIKING TIME – REF. 145.032

Combining a sonorous minute repeater with the contemporary visual appeal of two jumping-numeral apertures, the pink-gold Lange Zeitwerk Striking Time creates a multi-dimensional, multi-sensory spectacle of unobstructed clarity. In fact, the 44.2mm timepiece goes one step further, revealing the L043.2 caliber's polished steel hammers on either side of the 6:00 subsidiary seconds. The hammers, which indicate the quarter-hours and hours via high- or low-pitched tones respectively, may be silenced with ease by a push of a button at 4:00, preventing any disturbance if the wearer so desires. Apertures at 3:00 and 9:00 display the hours and minutes by way of three numeral discs that advance precisely with each passing minute thanks to the movement's patented constant-force escapement.

AUDEMARS PIGUET

MILLENARY MINUTE REPEATER – REF. 26371OR.OO.D803CR.01

The new Millenary Minute Repeater in rose gold is an exceptional, hand-wound wristwatch with an oval case and three-dimensional architecture that complements a refined and sophisticated movement highlighted by Audemars Piguet's own escapement, double balance spring, striking mechanism and gongs. The result is a visual and aural delight. The distinctively shaped case provides an instantly recognizable setting for the gold and enamel subdials, which invite closer inspection of the remarkable mechanism within. The manual-wound caliber is distinguished by the atypical construction of the regulating organ, which is composed of not just one balance spring, but of two placed top to tail.

BELL & ROSS

PW1 REPETITION 5 MINUTES SKELETON

Housed within a 52mm Argentium case, the PW1 Répétition 5 Minutes possesses an unusual complication, the five-minute repeater. This audible indication sounds, on request and musically, the hours and minutes thus: a low-pitched sound emitted for each hour, followed by a series of double low-pitched and high-pitched sounds for every five minutes elapsed. The complication is activated by pressing the button on the left-hand side of the case. The transparent dial, in conjunction with the sapphire crystal caseback, provides a full view of the manual-winding Dubois Dépraz movement with 56-hour power reserve. The caseback also bears a protective lid. Released in a limited edition of 25 pieces, the PW1 Répétition 5 Minutes is mounted on a silver chain.

BELL & ROSS

PW1 REPETITION MINUTES

Bell & Ross assigns new meaning to the pocketwatch, as its PW1 Répétition Minutes resonates with poetic melody. The watch keeps time like the clocks of yesteryear, as the wearer does not look for the hours to pass, but listens to them. Marked by sounds harmoniously emitted through its Argentium case, the PW1 Répétition Minutes has quite the complex configuration, indicating hours and minutes musically on request. To initiate the complication, the wearer presses a button on the left-hand side of the case. Upon activation, a low-pitched sound is emitted for the hours and a series of double low-pitched and high-pitched sounds indicates each increment of five minutes that have passed.

BLANCPAIN

CARROUSEL MINUTE REPEATER FLYBACK – REF. 2358-3631-55B

This unique model is truly a work of art. The masters at Blancpain include their well-known one-minute flying carrousel at 6:00. The minute repeater features a cathedral gong with blades that wrap one and a half times around the movement in order to emit a sound of surprising quality. The third function, which also stems from Blancpain's classic legacy, is a chronograph, whose pushbutton at 4:00 enables the wearer to restart a time measurement while a first measurement is in progress. This unique arrangement, featuring a 30-minute counter at the center of the caliber, demonstrates the horological mastery of the Blancpain movement design engineers.

CHRISTOPHE CLARET

ADAGIO – REF. MTR.SLB88.015-815

Adagio is an astounding minute repeater with cathedral gongs, featuring a large date at 6:00 and a second time zone between 1:00 and 2:00. As a tribute to his manufacture's first caliber introduced over 20 years ago, Christophe Claret has designed Adagio as a minute repeater—chiming the hours, the quarter-hours and minutes after the quarter-hours on demand—thus enabling the time to be heard rather than read. The in-house Caliber SLB88 is composed of 455 components. The cathedral gongs of the striking mechanism feature a patented device that prevents them from touching when they vibrate under the hammer strikes.

CHRISTOPHE CLARET

SOPRANO – REF. MTR.TRD98.020-028

The Soprano watch associates two of the finest horological complications: a 60-second tourbillon and a minute repeater with Westminster chime, four patented cathedral gongs and four hammers. The minute repeater mechanism is highlighted by stepped bridges inspired by the Charles X style, a favorite of Christophe Claret. This exceptional caliber is housed inside a round case subtly marrying precious metals and titanium. Tradition and modernity are united in this creation, entirely in keeping with the demands of ultra high-end watch making developed by the Manufacture Claret.

de GRISOGONO

OCCHIO RIPETIZIONE MINUTI – REF. OCCHIO S03

Housed in a blackened titanium case accented by 88 black diamonds on the bezel, this self-winding timepiece combines the classicism of a minute repeater with the contemporary ingenuity of an innovative shutter system. Upon the revelation of the movement via the slide-activated opening of the dial's 12-blade diaphragm, this monochrome wristwatch serenades its wearer with a three-chime cathedral gong that marks the hours, quarter-hours and minutes.

F.P. JOURNE

REPETITION SOUVERAINE

The ultra-slim F.P. Journe minute repeater presents an original construction to achieve its ultra-flat movement of barely 4mm. It includes 2 patents for the flat gongs that produce a louder and clearer sound, as well as an innovative system of striking racks and strike selection mechanisms which, placed under the dial, take up much less space than conventional strike mechanisms. The guilloché silver dial features a discreet window at 9:30 which features the hammers striking the gongs on the hour, the quarters and the minutes. The case is in steel for a better reverberation of the sound, with a guilloché silver dial and 18K rose-gold movement.

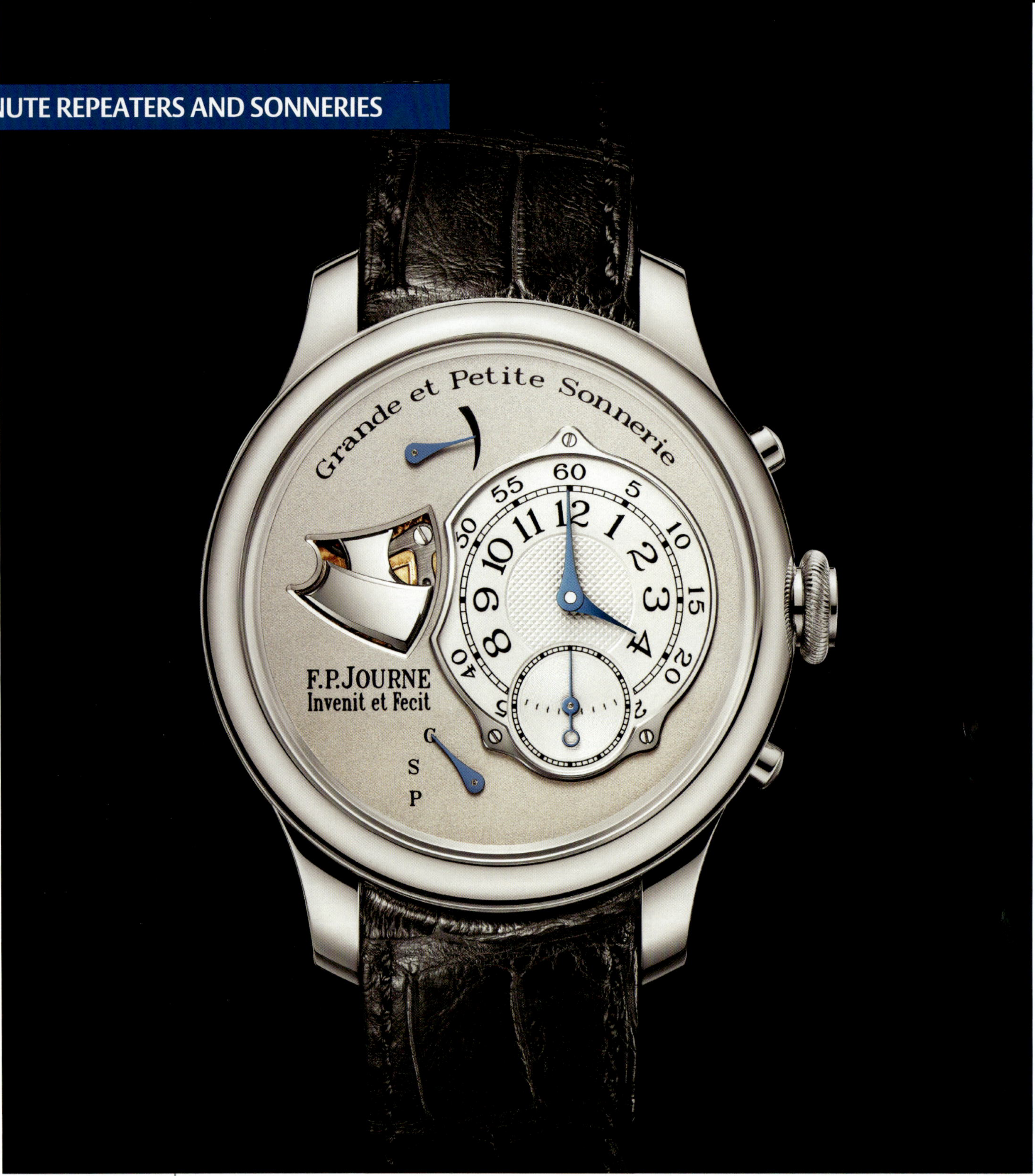

F.P. JOURNE

SONNERIE SOUVERAINE

The grande sonnerie with minute repeater wristwatch is one of the most complex horological creations. For the first time in history, a grand strike clock watch guarantees both functionality and total security offering true comfort, while 35,040 automatic chimes and 332,880 hammer strikes ring the time. The F.P. Journe Sonnerie Souveraine is protected by ten patents attesting to the "Invenit*" of this work of art, and six years of research and development to the "Fecit*" in the Manufacture's motto. Only four examples are produced by the F.P. Journe manufacture each year. The case is crafted in steel, with a white-gold and guilloché silver dial and an 18K rose-gold movement.

GUY ELLIA

REPETITION MINUTE ZEPHYR

The ambassador of Guy Ellia's complicated timekeeper, the Répétition Minute Zephyr with its complications and intricate manufacturing process, is considered a particularly exceptional piece. The Caliber GEC 88 was created by Swiss manufacture Christophe Claret and features a power reserve indicator, minute repeater function and five time zones with day/night indication. The movement beats at 18,000 vph in a solid, convex, high-resonance sapphire crystal case of impressive proportions (53.6x43.7x14.8mm). Mounted on an alligator strap with a folding buckle, this model is also available in pink gold or titanium.

JACOB & CO.

PALATIAL FLYING TOURBILLON MINUTE REPEATER – REF. 150.500.40.NS.OG.1NS

Beneath a semi-transparent dial in green mineral crystal, this limited edition wristwatch's hand-wound JCBM03 caliber drives a sophisticated timekeeping choreography. Boasting a titanium balance, one-minute flying tourbillon visible through the face at 6:00, and power reserve of 90 hours, the 308-component movement powers the sonorous delight of a minute repeater, activated via an 18K rose-gold sliding lever at 9:00. The 18K rose-gold case, measuring 43mm, provides a luxurious contrast to the dial's greenish tones with its adornment of a crown in 18K rose-gold.

PARMIGIANI

TORIC CAPITOLE – REF. PFH276-1203300-HA1441

Immediately standing out with its ingenious sectorial display of the hours and minutes, this 18K white-gold timepiece, with polished finish and fully hand-engraved and -knurled bezel, serenades its white mother-of-pearl dial with a two-gong "serpents" cathedral chime sounding the hours, quarter-hours, and minutes. Worn on an Hermès alligator leather strap, the watch is driven by the 397-component hand-wound PF 321 caliber with 45-hour power reserve. The demonstrative timepiece is finished with a sapphire crystal caseback engraved with the model's individual number.

PARMIGIANI

TORIC RESONANCE – REF. PFH.1200100.HA1441

Combining a minute repeater with a large date display is nothing new for Parmigiani, but the Toric Resonance represents a new feat for the brand, as it has never before offered instantaneous rotation of such a colossal date on this caliber. The discs on the date mechanism are the same size as the dial. Although they are incredibly finely cut, these discs require a lot of energy to activate them and even more to stop them dead following the instantaneous jump, without them rebounding under the effect of the forces exerted. This model has an 18K white-gold 45mm case and the minute repeater and other functions are displayed elegantly within.

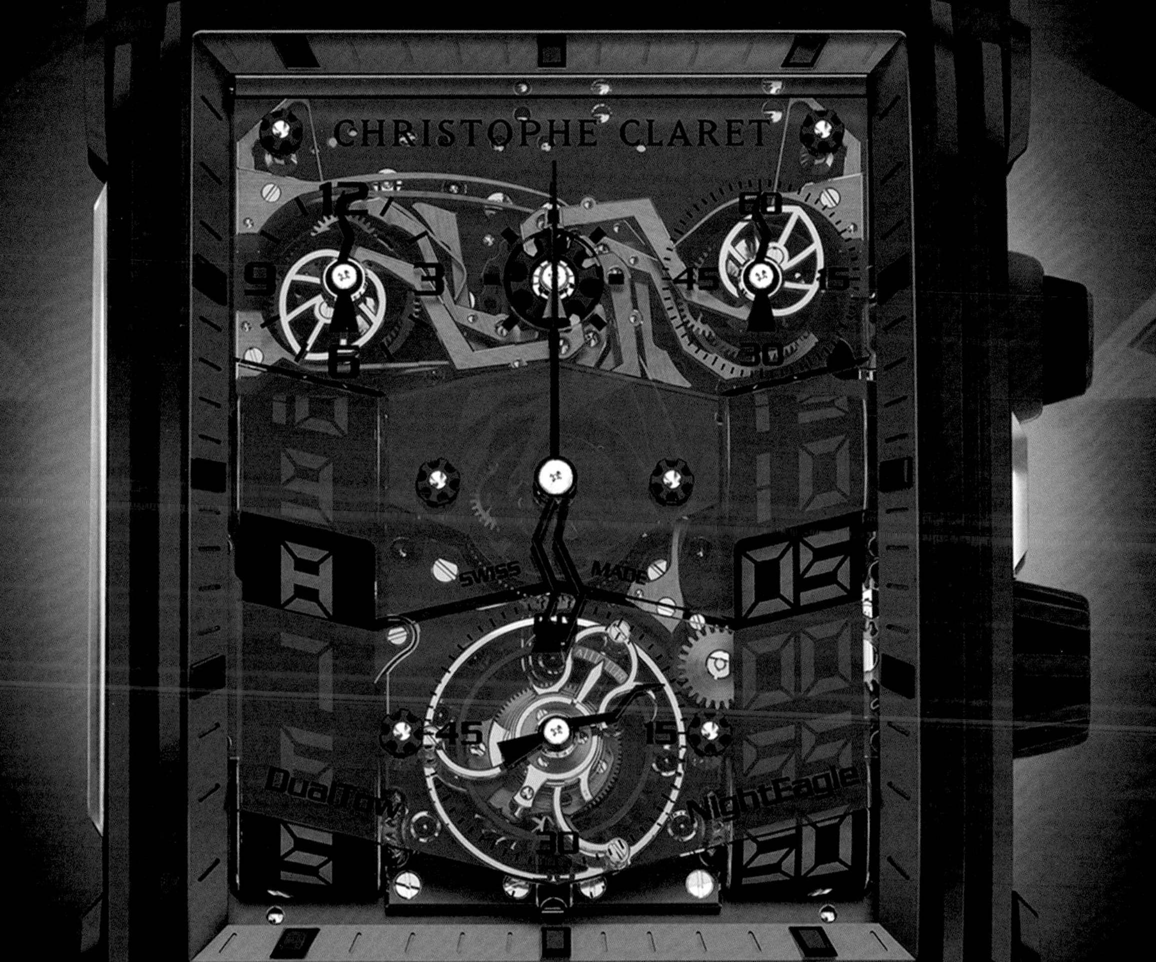
CHRISTOPHE CLARET
SWISS MADE
DualTow
NightEagle

Multi-Complications

Hybris Mechanica à Gyrotourbillon

Reverso à Tryptique Hybris Mechanica

Jaeger-LeCoultre, *Hybris Mechanica*

The 2009 Hybris Mechanica trilogy from Jaeger-LeCoultre displays a total of 55 complications, some of which are found on more than one model. The Hybris Mechanica à Gyrotourbillon is an homage to the quintessence of the mechanical measure of time, thanks to its "spherical tourbillon" (see chapter on Tourbillons), completed by a perpetual calendar with four retrograde hands and an equation of time function that can be adjusted according to one's location. The Hybris Mechanica à Tryptique comes up with an unusual take on the famous Reverso reversible case, originally created in 1931. The watch boasts three faces, for three dimensions of time. On the front of the case is civil time, with a new detent escapement system and a titanium tourbillon; the back shows sidereal time, with a sky map, Zodiac calendar, equation of time and times of sunrise and sunset; while the watch cradle displays perpetual time, with a complete calendar—making a grand total of 18 complications on this exceptional timepiece. As for the Hybris Mechanica à Grande Sonnerie, it contains no fewer than 26 complications, with a 1,300-piece movement that powers a flying tourbillon, perpetual calendar with retrograde hands and a strike/minute repeater mechanism with numerous patented innovations.

Hybris Mechanica à Grande Sonnerie

Dominique Loiseau, *1F4*

Produced in just two models per year, the 1F4 from Dominique Loiseau associates numerous functions, including a perpetual calendar, split-seconds chronograph, second time zone, equation of time, flying tourbillon, minute repeater and grand and small strike.

Franck Muller, *Aeternitas Mega 4*

The Aeternitas Mega 4 model from Franck Muller houses, within its elegant tonneau-shaped case, 15 complications, including grand and small strike, minute repeater with Westminster chime, secular calendar with retrograde date, equation of time, large tourbillon, split-seconds chronograph and two additional time zones.

Vacheron Constantin, *Tour de l'Ile*

In 2005, Vacheron Constantin celebrated its 250th anniversary by creating the ultra-complicated model called Tour de l'Ile. This double-sided timepiece contains in its 834-piece movement no fewer than 16 complications, including a minute repeater, tourbillon, indications of the moonphase and age of the moon, perpetual calendar, equation of time, times of sunrise and sunset, and a map of the night sky.

Patek Philippe, *Grandmaster Chime*

Created to celebrate the 175th anniversary of Patek Philippe (founded in 1839), the Grandmaster Chime combines a total of 20 complications, including a large and small strike, a minute repeater, an instantaneous perpetual calendar and two patented innovations: an acoustic alarm that chimes the time of the alarm and a date repeater that chimes the date on demand. Its 1,366-piece movement is housed in an entirely hand-engraved, double-face case, complete with an ingenious reversing mechanism integrated in the lugs.

Varied Triumphs

The horological superstars whose complications can be counted on the fingers of two, three, four, even five or six hands should not make us forget all the creations that—though they have achieved more modest recognition—are exceptional triumphs in their own right. A "grand complication," in the watch world, is created when a watch contains three or four supplementary functions—especially if each of those functions belongs to one of the traditional domains of haute horology (audible indications, calendars, chronographs, tourbillons, etc.).

A. Lange & Söhne, *Grand Complication*
The Grand Complication model is the most sophisticated watch from A. Lange & Söhne and combines a large and small strike (with minute repeater), a perpetual calendar and a split-second chronograph with jumping seconds. It is issued in a six-piece limited edition.

Blancpain
Le Brassus Carrousel Répétition Minutes Chronographe Flyback
On the Le Brassus Carrousel Répétition Minutes Chronographe Flyback model from Blancpain, the openworked dial and sapphire exhibition caseback provide a fine opportunity to admire the triple complication in action—and notably the famous carrousel (or karussel), an alternative to the tourbillon that was revived by the brand in 2008.

Audemars Piguet, *Royal Oak Offshore Grande Complication*

Equipped with highly visible architecture that reveals its openworked movement, the Royal Oak Offshore Grande Complication brings together a perpetual calendar, minute repeater and split-seconds chronograph within its titanium and ceramic case.

Vacheron Constantin
Patrimony Traditionnelle Calibre 2755

A worthy descendant of the prestigious Tour de l'Ile, released for the 250th anniversary of the brand, the Patrimony Traditionnelle Calibre 2755 from Vacheron Constantin brings together three major complications of the horological universe—minute repeater, tourbillon and perpetual calendar—in an elegant round case.

Audemars Piguet
Tradition Minute Repeater Tourbillon Chronograph

The Tradition Minute Repeater Tourbillon Chronograph by Audemars Piguet accommodates its three complications within a cushion-shaped pink-gold or titanium case topped by a white-gold bezel. A transparent exhibition caseback provides a chance to admire the movement decorations.

IWC, *Portuguese Grand Complication*

The Portuguese Grand Complication from IWC offers a minute repeater and chronograph, as well as a perpetual calendar that indicates not only the date, day and month, but the year as well, via a four-digit display.

Breguet
Classique Grand Complication Skeleton Tourbillon Perpetual Calendar

The Classique Grand Complication Skeleton Tourbillon Perpetual Calendar from Breguet stages its two complications within a completely transparent architecture, showcasing an entirely skeletonized movement.

Patek Philippe, *Grand Complications Ref. 5216*

Reinterpreting one of the manufacture's classics, Patek Philippe's Grand Complications Ref. 5216 conceals beneath its sober dial a minute repeater, tourbillon, perpetual calendar with retrograde date hand, and a moonphase display.

Jean Dunand, *Shabaka*

Boasting a highly original, Art Deco-inspired dial, Shabaka from Jean Dunand associates a minute repeater with cathedral chime with a complete perpetual calendar with instantaneous day/date/month displays on rollers.

Patek Philippe
Triple Complication Ref. 5208
The automatic Triple Complication Ref. 5208 wristwatch from Patek Philippe associates within its platinum case a minute repeater, single-pusher chronograph and instantaneous perpetual calendar with apertures. It also stands out for its escapement and balance spring in Silinvar®, a silicon derivative.

Jean Dunand, *Grande Complication*
In 2005, the brand Jean Dunand—co-founded by Christophe Claret, the renowned constructor of complicated movements—presented a wristwatch called Grande Complication. This 827-piece timepiece includes, among other functions, a single-pusher split-seconds chronograph, tourbillon, minute repeater and perpetual calendar with retrograde date and day indications.

Maîtres du Temps, *Chapter One*
Presented in 2008 in a tonneau-shaped case and rereleased in 2010 in a round version, Chapter One from Maîtres du Temps unites a single-pusher column-wheel chronograph, a retrograde second time zone display and retrograde date, as well as day and moonphase displays on two rollers at six o'clock and twelve o'clock.

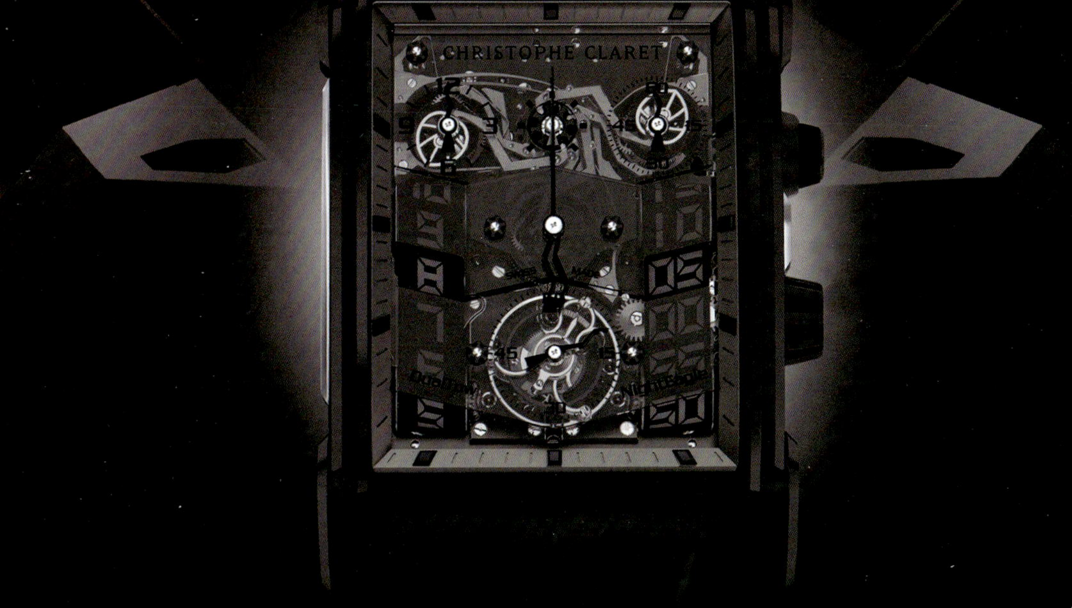

Christophe Claret, *DualTow NightEagle*

Christophe Claret scored a huge coup with its DualTow NightEagle, which features hour/minute displays on two belt drives inspired by treaded vehicles, a single-pusher planetary chronograph with strike and tourbillon, not to mention many other refinements and technical innovations.

HD3, *Slyde*

An ultra-modern, ultra-geeky version of grand complications, the new electronic model Slyde from HD3 is distinguished by its curved tactile screen that displays the various indications (time, calendar, chronograph, countdown clock, moonphase, second time zone) in much the same way as a smartphone.

A. LANGE & SÖHNE

LANGE 1 TOURBILLON PERPETUAL CALENDAR – REF. 720.025

The Lange 1 Tourbillon Perpetual Calendar's self-winding L082.1 caliber is ingeniously fitted with a platinum centrifugal mass for exceptional winding efficiency. Thanks to an innovative rotating peripheral month ring that occupies a minimal amount of space, and an hour and minute circle off-centered on the solid silver rhodiumed dial, A. Lange & Söhne accomplishes a perpetual calendar of remarkable ease of use and legibility, despite its comprehensive constitution of moonphase, retrograde day, leap-year aperture and date. Admired intimately through the sapphire crystal caseback of this 41.9mm platinum masterpiece, the exquisite tourbillon is embellished with an extremely sophisticated black polished finish, and boasts a flawless diamond end stone as its radiant bearing.

A. LANGE & SÖHNE

RICHARD LANGE PERPETUAL CALENDAR "TERRALUNA" – REF. 180.026

This hand-wound wristwatch houses its 787-component L096.1 caliber in a 45.5mm white-gold case. The movement, endowed with a sophisticated constant-force escapement that ensures unvarying amplitude throughout the 14-day power reserve, animates a perpetual calendar of ingenious architecture. Told on three separate axes, the hours, minutes and seconds are met in their respective subdials by apertures displaying the day, month and large date. A discreet leap-year indicator at 2:30 completes the dial-side exhibition. On the caseback, the Terraluna's orbital moonphase display depicts the moon orbiting the earth once per precise synodic month while the moon's phases and relative position are illustrated via a constellation view that incorporates the balance as a representation of the sun.

AUDEMARS PIGUET

JULES AUDEMARS CHRONOMETER WITH AP ESCAPEMENT – REF. 26153OR.OO.D088CR.01

This hand-wound, COSC-certified chronometer boasts a power reserve of 90 hours with parallel twin barrels. The high-frequency model oscillates at 42,300 vph and is framed by an 18K pink-gold case with a diameter of 46mm. The off-centered hour and minute counter at 12:00 is in black enamel, as is the small seconds counter at 6:00. Both counters contain white printed Roman numerals and pink-gold hands. This eye-catching timepiece is available on a brown hand-stitched large square scale alligator leather strap with an 18K pink-gold AP folding clasp.

AUDEMARS PIGUET

OPENWORKED ROYAL OAK OFFSHORE GRANDE COMPLICATION REF. 26571IO.OO.A010CA.01

The automatic-winding openworked Manufacture caliber 2885 powers this intricately crafted timepiece. The perpetual calendar indicates the day, date, week, month, leap year and moonphase alongside a chronograph with a small seconds counter at 9:00, and the complications don't stop there. The minute repeater's two gongs add an aural element to the visually striking model. The wearer can see the complex movement through the smoked sapphire dial.

AUDEMARS PIGUET

ROYAL OAK CONCEPT GMT TOURBILLON – REF. 26580IO.OO.D010CA.01

Bright, white, sleek and powerful, the Royal Oak Concept GMT Tourbillon is sculpted with a titanium case middle and integrated rubber strap. The white ceramic bezel frames the highly architectural array of tourbillon carriage and second GMT time zone display within. The white center stage, an intricately machined upper bridge made of white ceramic, matches the ceramic in the movement itself, visually transforming the new caliber 2930. The tourbillon cage comprises 85 components but is only 0.45g in weight. Each component is beveled, polished, assembled and polished by hand. The GMT display provides an instant reading of the time in a second time zone. It is adjusted using the pushbutton at 4:00.

AUDEMARS PIGUET

ROYAL OAK OFFSHORE TOURBILLON CHRONOGRAPH – REF. 26387IO.OO.D002CA.01

Available in a limited edition of 30 models, this collector's item is designed for those who dare to be different. The manual-winding Manufacture caliber 2933 has hand-crafted finishing on the bridges, mainplate and cut-out parts that are visible to the owner. The silver-toned dial is openworked on the tourbillon at 9:00 and also at 6:00 and 12:00. The chronograph is complete with a small seconds counter at 3:00. Large, luminous Arabic numerals and blackened gold Royal Oak hands complete the dial. The black rubber strap and titanium AP folding clasp reaffirm the strength of this special piece.

AUDEMARS PIGUET

ROYAL OAK OFFSHORE TOURBILLON CHRONOGRAPH – REF. 26550AU.OO.A002CA.01

This piece unites the tourbillon and chronograph in a harmonious fashion and is available in a limited edition of 50 pieces. The black dial is complete with the signature Audemars Piguet "Mega Tapisserie" pattern, black chronograph counters, gold applied hour markers and gold Royal Oak hands with luminescent coating. The tourbillon stands prominently at 6:00. Powered by the automatic-winding manufacture caliber 2897, this model oscillates at 21,600 vph and has a power reserve of 65 hours.

BLANCPAIN

VILLERET CARROUSEL PHASES DE LUNE – REF. 6622L-3631-55B

The new Blancpain 225L movement unites a carrousel, a moonphase and a date for the very first time. The carrousel was brought back by Blancpain to the forefront of the watchmaking scene in 2008, with the introduction of the world's first flying carrousel to perform a one-minute rotation. The movement of the Villeret Carrousel Phases de Lune is equipped with silicon balance-spring technology and endowed with a five-day power reserve. The double-stepped 42mm red-gold case frames a cambered Grand Feu enamel dial; a blued steel serpentine hand indicates the date, while the hour and minute hands sweep across painted enamel Roman numerals. The white dial features a 12:00 opening revealing the carrousel carriage with its complex mechanism, set exactly opposite the moonphase indication at 6:00.

BREGUET

CLASSIQUE GRANDE COMPLICATION SKELETON TOURBILLON PERPETUAL CALENDAR REF. 3795BR-1E-9WU

Boasting a tourbillon and perpetual calendar within its rose-gold case's 41mm diameter (a platinum version is also available), the Classique Grande Complication Skeleton Tourbillon Perpetual Calendar presents its visually striking complications within a complex yet legible display, as well as through a sapphire crystal caseback. A skeletonized dial in silvered 18-karat gold provides a stage for: an hour chapter with Roman numerals on a sapphire crystal disc; retrograde date indication at 12:00; months and leap years at 3:00; small seconds display and tourbillon at 6:00; and days at 9:00. The manual-winding movement, Cal. 558QP3, measures 14 ½ lines, uses 21 jewels and disposes of a power reserve of 50 hours.

BREGUET

CLASSIQUE GRANDE COMPLICATION TOURBILLON PERPETUAL CALENDAR REF. 3797BR-1E-9WU

The dial of the Classique Grande Complication Tourbillon Perpetual Calendar strikes an elegant note: crafted in 18-karat gold, silvered and engine-turned in four different patterns, it reveals the tourbillon at 6:00 and Breguet's care with the most precise details. The 41mm platinum case is girded by a finely fluted caseband. Complementing the tourbillon are the indications arrayed around the watch face: retrograde date indication and an hours chapter with Roman numerals on a sapphire disc at 12:00, months and leap years at 3:00, small seconds on the tourbillon axis at 6:00 and days at 9:00. The manually wound movement, Cal. 558QP2, provides 50 hours of power reserve and can be seen through a sapphire crystal caseback. This timepiece is also available in rose gold.

CHRISTOPHE CLARET

BACCARA – REF. MTR.BCR09.020-029

Baccara is a timepiece doubling as a miniature casino with three games: baccara, roulette and dice, each appealing to the audio, visual, and tactile senses. The Baccara cards appear on the dial as if by magic. At 6:00, the player's cards are distributed in three small windows. The banker's cards are at 12:00. To shuffle the cards, the player presses a pusher located at 9:00 while a pusher at 8:00 distributes the cards to the players. At 10:00, a third pusher organizes distribution to the bank. Every time the shutter opens, for either the player or the bank, a cathedral gong can be seen through a caseband window at 2:00.

CHRISTOPHE CLARET

BLACKJACK – REF. MTR.BLJ08.340-361

A veritable miniature casino, the Blackjack matches grand complications with the world of gaming, in the process creating a new watchmaking paradigm: the interactive watch. Rather than just tell the time, the wearer can enjoy the sensory effects of blackjack, roulette and dice. On the lower part of the dial, the player's four cards appear in windows. Two are visible, the other two hidden by shutters. On the upper part of the dial are three additional windows for the dealer's card, one of which is visible, the other two also hidden by shutters. The Blackjack is an unprecedented upmarket timepiece for aficionados, initiating a kind of watchmaking that has cast off its inhibitions.

CHRISTOPHE CLARET

MAESTOSO – REF. MTR.DTC07.030-050

With Maestoso, Christophe Claret took up the challenge of incorporating a traditional pivoted detent escapement, usually designed to run in a perfectly stabilized position, in a wristwatch. Enabling the detent escapement to operate in all the positions that wearing a wristwatch entails is a genuine gamble, pulled off via three patented mechanisms and a series of additional, innovative systems. To prevent the detent from turning over, an anti-pivot cam, integral to the spring balance, works in conjunction with the safety finger. The ensemble is fitted between a mainplate and two sapphire bridges and pivots on a ball bearing, a patented mechanism, that distributes the load on the escapement so it de facto absorbs impacts by means of a spring, which gives it the requisite flexibility.

CHRISTOPHE CLARET

MARGOT – REF. MTR.EMT17.000-020

Christophe Claret presents his first watch for women with the unique wonder that is the Margot. The inventive complication pays homage to the classic game that daisy-picking, hopeless romantics have played since the Middle Ages: "He loves me...he loves me not." In the French version of this age-old game, there are far more options: he loves me a little, loves me a lot, loves me passionately, loves me madly, or doesn't love me at all. These possibilities are featured on the ultra-innovative timepiece as the wearer presses the pusher at 2:00 to answer the question: "Does he love me?" The Margot is available in four limited editions of 20 pieces each.

CHRISTOPHE CLARET

POKER – REF. MTR.PCK05.001-020

Poker is the very first timepiece that manages to replicate the card game in an automaton watch. While the game seems simple to organize at a table, its watchmaking solution is an extreme test and no technical solution had been found until Mr. Christophe Claret himself came up with the answer. While the first prototype was operational in 2011, it took two years to perfect the intricate complication. In total, Poker packs 32,768 different combinations, i.e. 98,304 combinations for three players. The probabilities were calculated so that each player has approximately the same chances of winning. The original automatic-winding PCK05 caliber comprises 655 components and has a power reserve of 72 hours. There are bound to be some late nights! In addition to its gaming functions, Poker has an extremely legible display of hours and minutes.

DE BETHUNE

DB16 TOURBILLON REGULATOR PLATINUM – REF. DB16PS2

Housed within a 43mm platinum case with cone-shaped lugs, De Bethune's DB16 Tourbillon Regulator Platinum uses its 5N rose-gold dial, adorned with a sunburst hand-guilloché motif, to indicate the hours, minutes, jumping seconds, day, date, month, spherical moonphase and leap year indication. The sapphire crystal caseback reveals a stunning spectacle: the hand-wound DB2509 caliber, which powers a silicon/titanium tourbillon with 30-second indication, central jumping seconds mechanism, retrograde age of moon indication and power reserve indication, all visible from the back. Powering this multi-complicated watch at a frequency of 36,000 vph, the DB2509 caliber possesses 48 jewels and boasts such De Bethune innovations and registered patents as a self-regulating twin barrel, a silicon/white-gold balance wheel and balance-spring with flat terminal curve.

DE BETHUNE

DB25 LT TOURBILLON – REF. DB25LTP

De Bethune consistently pursues a balance between simplicity, readability and precision. Crafted in this spirit, the brand's 44mm DB25 LT Tourbillon, built on a set of superposed rings of various sizes, combines the cool tones of platinum with 5N rose gold and a hand-guilloché dial. The spherical De Bethune moon sits at 12:00; composed of half-spheres set into blued steel and palladium, this three-dimensional indicator is driven by a high-precision mechanism requiring only one adjustment every 1,112 years. Alongside classic features such as a triangular bridge adorned with Côtes de Bethune, the caseback displays an avant-garde mechanism powered by a 30-second silicon and titanium tourbillon whose balance-wheel oscillates at 36,000 vph, making it one of the fastest and lightest in the watch industry with 63 components weighing just 0.18g in all.

DE BETHUNE

DB29 MAXICHRONO TOURBILLON – REF. DB29RS1

The DB29 Maxichrono Tourbillon, the result of seven years of research and development, is one of a kind. The rose-gold case of the DB29 features the brand's trademark cone-shaped lugs. A single pushbutton at 3:00 operates the chronograph. The five-hand design of the dial makes it easier to read the chronograph indications by doing away with the conventional subdials. Thanks to the precision ensured by the De Bethune 30-second tourbillon oscillating at a rate of 36,000 vph, the maximum elapsed time measurement amounts to a generous 23 hours, 59 minutes and 59.9 seconds. A pushbutton at 4:00 opens to reveal through a sapphire crystal the modern construction of the DB2039 movement with polished steel bridges, tourbillon at 3:00 and the complex mechanism of an innovative time-measuring system featuring three column wheels.

GIRARD-PERREGAUX

CONSTANT ESCAPEMENT L.M. – REF. 93500-53-131-BA6C

A true technical revolution that stunned connoisseurs when prototypes of the innovative mechanism were first presented, the Constant Escapement is now integrated into the movements driving new models of Girard-Perregaux's Haute Horlogerie collection. Based on groundbreaking prototypes presented in 2008, it took five years of research and development to bring this signature movement to fruition. Unlike other systems that propose a constant force averaged over time, this is an authentic constant-force escapement, as the force is both instantaneous and continuous. This is the first model to house this innovative and original escapement, and it is garbed in resolutely technical and contemporary design.

GIRARD-PERREGAUX

CONSTANT ESCAPEMENT L.M. PINK GOLD – REF. 93500-52-731-BA6D

To accomplish the watchmaking dream of constant force, Girard-Perregaux watch masters have integrated an intermediate device into the escapement, consisting of an extremely thin silicon blade that stores the energy up to an invariable threshold bordering on instability, and then transmits it all instantaneously, before recommencing the cycle. This groundbreaking timepiece is now available in a pink-gold case with a softly domed, varnished white dial that exudes timelessness and elegance in every respect. The warm hue of the 18K pink gold version compliments the traditional architecture and intricacies within this masterpiece of watch mechanics. The faceted dauphine hands and applied hour markers have also been redesigned in the same spirit and color. Despite its 48mm diameter, the sleek case with softly curved lugs gently hugs the wrist for perfect comfort.

HUBLOT

CLASSIC FUSION TOURBILLON CATHEDRAL MINUTE REPEATER – REF. 504.NX.0170.LR

The sapphire dial of Hublot's Classic Fusion Tourbillon Cathedral Minute Repeater reveals the Manufacture-made HUB 8001 hand-wound mechanical movement, with a 13.6mm tourbillon cage and a minute repeater mechanism. The assembly of its 319 components is a unique decorative element in itself, beautifully complemented by the watch's hands, which are rhodium-plated to visually echo the 45mm titanium case. One of the technical feats achieved on the minute repeater is the complete integration of the trigger piece in the left-hand bezel lug of the Classic Fusion's case. Finally, the timepiece, which has a power reserve of five days, is water resistant to 30m—a highly complex achievement. The sound of the chime no longer relies on airborne transmission, but rather resonates through the material of the case itself.

IWC

AQUATIMER CHRONOGRAPH EDITION "EXPEDITION JACQUES-YVES COUSTEAU" REF. IW376805

The Aquatimer Chronograph Edition "Expedition Jacques-Yves Cousteau" is a memento of Cousteau's voyage of discovery to the Galapagos archipelago. With Jacques Cousteau, IWC Schaffhausen shares not only a technological pioneering spirit and a passion for perfection, but also the commitment to an intact environment worth living in. This 44mm mechanical self-winding chronograph boasts a 44-hour power reserve and a day and date display at 3:00. An Aquatimer Chronograph through and through, it boasts all the features of a diver's watch designed for expedition use. The engraving on the back of the cover shows "le Commandant" with his trademark red woolen beanie.

IWC

AQUATIMER PERPETUAL CALENDAR DIGITAL DATE-MONTH – REF. IW 379401

Available in a limited edition of 50 pieces, the Aquatimer Perpetual Calendar Digital Date-Month is the flagship of the Aquatimer collection. The dial is distinguished by the extra-large numerals of the perpetual calendar, mechanically programmed until 2100. The date and month discs have perforated covers, inspired by the filter systems that are omnipresent on submarine vehicles, that provide a view of the complex switching mechanism at work. The four-year leap year cycle is also shown digitally. The chronograph display gives the diver a convenient way of sticking to the prescribed surface interval time between one dive and the next. A subdial at 12:00 displays stop times between one minute and 12 hours as simply as the time of day, using minute and hour hands, while the central hand counts the number of seconds.

JACOB & CO.

ASTRONOMIA TOURBILLON –REF. 750.100.94.AB.SD.1NS

Released to great acclaim at BaselWorld 2014, The Astronomia Tourbillon is both a visual marvel and a technical triumph for the masters at Jacob & Co. Alluding to the celestial world, this poetic model's heart is a four-armed carrier which rotates every twenty minutes. On one arm lies a triple-axis gravitational tourbillon, which oscillates at 18,000 vph. The openworked dial allows the wearer to see the rotating 56-facet diamond moon in addition to the rotating Earth in 18K white gold with enamel.

JAEGER-LECOULTRE

HYBRIS MECHANICA 11 – REF. Q1313520

Hybris Mechanica 11, also known as the Master Ultra Thin Minute Repeater Flying Tourbillon, is a true trailblazer in the field of ultra-thin grand complication timepieces. This model is equipped with the very latest technological innovations developed by the Manufacture in the field of ultra-thin timepieces. These technical advances enable it to break a new record for thinness. With a thickness of only 7.9mm, the Hybris Mechanica 11 can claim its place as the thinnest of all minute repeater watches. In addition to the minute repeater function, it also features for the first time a flying tourbillon combined with a flying balance and a peripheral self-winding system.

JAEGER-LECOULTRE

MASTER GRANDE TRADITION TOURBILLON CYLINDRIQUE A QUANTIEME PERPETUEL REF. Q5042520

This exceptional watch is a perfect combination of the inspirational influence of the historic timepieces of the 19th century and today's cutting-edge technology. Housed in an 18K pink gold case, this watch's dial displays the indications of the perpetual calendar function in an ultra-legible way. Equipped with a Jaeger-LeCoultre 985 self-winding movement, it combines a flying tourbillon and a cylindrical balance spring that guarantees exceptional timekeeping performance. In short, this is a timepiece of great precision that combines exceptional complications and an extremely expressive aesthetic style.

PARMIGIANI

TONDAGRAPHE – REF. 236.1000300.HA1441

Originally introduced in 2009, the Tondagraphe later added two outstanding features: a grand complication combining a chronograph and a tourbillon at 6:00, housed in a sleek rose-gold case in the Swiss manufacture's signature style. This model is brought to life by a house-made caliber, which boasts a power reserve of 72 hours and combines the intricate beauty of the tourbillon with the technical expertise of the chronograph. The expertly crafted tourbillon stands out at 6:00 and the cage that encloses the regulating organ turns on its axis at a rate of one rotation per minute. The dial has enhanced legibility with the addition of elegant Parmigiani numerals.

PARMIGIANI

TORIC TECNICA OMBRE BLANCHE – REF. PFH466.1002400.HA1241

Powered by the hand-wound manufacture made PF351 caliber, the Tecnica is comprised of 504 parts and 39 jewels with a power reserve of 42 hours. The dial features Grand Feu enameling and is housed in a 45mm 18K rose-gold case. The case is very thin, measuring a mere 13mm, quite impressive considering all of the complications housed within, especially the minute repeater, which typically calls for a thicker construction. The minute repeater is wound by rotating a knurled ring under the bezel a quarter-turn clockwise, ultimately giving the best possible resonance in an optimal space. Perpetual calendar indications surround the center of the dial, and a tourbillon whirls at 6:00.

PATEK PHILIPPE

MEN GRAND COMPLICATIONS – REF. 5204P-011

Upon a chic ebony black opaline dial, this Men Grand Complications model, housed within a 40mm platinum case, features a plethora of indications. At 12:00, the day and month appear through two apertures, with date and moonphase in a subdial at 6:00. The leap year cycle and day/night indication round out the calendar displays. The split-seconds chronograph operates with two central hands and a 30-minute calendar at 3:00. Small seconds are located at 9:00. The manual-winding CHR 29-535 PS Q movement within beats at 28,800 vph, with a power reserve of 55-65 hours. Its 496 components and 34 jewels fit together perfectly within its 32mm diameter and 8.7mm thickness, and it has passed the stringent requirement to earn the Patek Philippe Hallmark.

PATEK PHILIPPE

MEN GRAND COMPLICATIONS – REF. 5217P-001

A bezel and lugs set with 56 baguette diamonds (approximately 3.54 carats) add an extra sparkle to this elegant multi-complicated timepiece. The 39.5mm platinum case bears a slide piece on the side to activate the piece's two-gong minute repeater function. The black lacquer dial sports diamond hour markers and a full complement of perpetual calendar indications: day and month through two apertures on either side of the dial, leap year cycle through an aperture at 12:00, moonphase (and small seconds) in a subdial at 6:00, and a red-tipped hand that swings across a retrograde arc to indicate the date. The tourbillon at 6:00 is incredibly subtle, noted with just the complication's name on the dial. Endowed with the Patek Philippe Hallmark, this highly complicated model is mounted on a shiny black alligator leather strap.

PATEK PHILIPPE

MEN GRAND COMPLICATIONS – REF. 5270G-013

This 41mm white-gold Men Grand Complications timepiece, endowed with the Patek Philippe Hallmark, combines the attributes of a perpetual calendar and chronograph, with a tachometer scale to boot. Displaying day and month at 12:00, leap year cycle at 4:30, date and moonphase at 6:00 and day/night indication at 7:30, the model also includes a 30-minute chronograph counter with central chronograph seconds and a tachometer scale around the dial circumference. The manual-winding CH 29-535 PS Q caliber, with 456 components and 33 jewels, provides 55-65 hours of power reserve.

PATEK PHILIPPE

MEN GRAND COMPLICATIONS – REF. 5271P-001

Upon a black-lacquer dial with gold hour marker appliqués, this Men Grand Complications, endowed with the Patek Philippe Hallmark, displays all the perpetual calendar indications, including day and month (through apertures at 12:00), leap year cycle (at 4:30), date and moonphase (at 6:00). A 30-minute counter (at 3:00), day/night indication (at 7:30) and small seconds display (at 9:00) round out the dial displays, girded by a tachometer scale. Fifty-eight baguette diamonds adorn the bezel and lugs of the 41mm platinum case. The hand-wound CH 29-535 PS Q caliber within drives all indications with a power reserve of 55-65 hours.

PATEK PHILIPPE

MEN GRAND COMPLICATIONS – REF. 5304R-001

Presenting a splendid display through the sapphire crystal dial, this Men Grand Complications timepiece, endowed with the Patek Philippe Hallmark, uses a 43mm rose-gold case to house the automatic-winding R 27 PS QR LU caliber. With a diameter of 28mm and a thickness of 7.23mm, this movement is composed of 517 components and 41 jewels, and possesses a power reserve of 38-48 hours. This exceptional timepiece features the following complications: minute repeater, perpetual calendar with retrograde date hand and day, month and leap year apertures, moonphase and age of moon. The crescent-tipped date hand swings around an arc at the very circumference of the dial, upon a black ring that matches the strap, which is hand-stitched in shiny black alligator leather.

PIAGET

PIAGET GOUVERNEUR – REF. G0A39116

Equipped with the new ultra-thin manual-wound 642P moonphase tourbillon movement, this model exemplifies a true feat of haute horology. It is extremely slim, measuring just 4mm in thickness, and also contains two prestigious grand complications. The moonphase is at 6:00 while the flying tourbillon at 12:00 contains a titanium carriage and small seconds. The hours and minutes are central. The Gouverneur is not just a horological masterpiece, but also an aesthetic wonder, as the 43mm rose-gold case is enhanced by two rows of delicately set diamonds.

PIAGET

PIAGET GOUVERNEUR TOURBILLON – REF. G0A37115

A member of Piaget's Black Tie collection, the Gouverneur Tourbillon combines elegant aesthetics with technical expertise. Entirely manufactured in house, this model is equipped with the ultra-thin, hand-wound 642P caliber beating at 21,600 vph with a power reserve of 40 hours. The silvered guilloché dial showcases a tourbillon with small seconds on the cage at 12:00, as well as an astronomic moonphase at 6:00. The silvery, cool color of the dial is complemented by the 18K white-gold case, which is perfectly set with two rows of diamonds (1.4 carats). This model possesses two of the grand complications and embodies elegance and beauty.

ROGER DUBUIS

EXCALIBUR QUATUOR IN PINK GOLD – REF. RDDBEX0367

The Excalibur Quatuor watch is produced in a limited series. It represents innovation while respecting the great traditions of watchmaking by incorporating four balance springs as well as a completely original power reserve indicator. These two technical feats have not been developed for their own sake but as part of watchmaking's permanent pursuit of precision. The precision of a movement is affected by the gravity of the Earth as the watch constantly changes position with the motion of the wrist. While the tourbillon has provided a partial answer to this problem, the Excalibur Quatuor offers a new solution, with its four carefully positioned sprung balances working in pairs. This immediately compensates for the rate variations caused by changes in position.

TAG HEUER

CARRERA MIKROPENDULUMS – REF. CAR.2B83.FC6339

An exclusive TAG Heuer-patented technology: the Carrera MikropendulumS contains the first magnetic double tourbillon. Equipped with two magnetic pendulums that replace traditional hairsprings, this chronograph tourbillon rotates 12 times per minute. The two tourbillon pendulums and their solid 5N 18K rose-gold bridges are visible through the fine-brushed anthracite dial. The case is forged from a revolutionary material, a chrome and cobalt alloy used in aviation and surgery. It is fully biocompatible, harder than titanium, easier to shape and as luminous as white gold. The case design, with its stopwatch-like placement of the crown at 12:00, is based on the 2012 Aiguille d'Or winner, the TAG Heuer Carrera Mikrogirder, and the Carrera 50 Years Anniversary Jack Heuer edition.

TAG HEUER

CARRERA MIKROPENDULUM TOURBILLON – REF. CAR5B81.FC6331

TAG Heuer has made a leap in avant-garde tourbillon technology with the Carrera Mikropendulum Tourbillon. This timepiece is the first to combine a mechanical tourbillon for the watch with a magnetic tourbillon for the chronograph. This groundbreaking regulating system is accurate to 1/100th of a second, resulting in one of the most daring and precise complications ever created. Although the tourbillon is the most iconic and traditional horological complication, there is nothing conventional about this piece. Comprised of 454 components, the movement has two barrels as energy sources and two regulating systems that eliminate the need for a clutch, as both the piece in its entirety and the chronograph have their own independent sources of energy.

ZENITH

ACADEMY CHRISTOPHE COLOMB FELIX BAUMGARTNER – REF. 39.2210.8804/58.C714

This piece is a tribute to Felix Baumgartner's record-setting free fall from an altitude of 38,969.4m. The watch boasts a groundbreaking gyroscopic module, a first in mechanical horology, as the centerpiece of its design. The self-regulating "gravity control" mechanism ensures the constant horizontal positioning of the escapement thanks to an award-winning innovation inspired by marine chronometry's gimbal suspension systems. This complex system pierces the bright blue, turquoise and aventurine dial. Housed within the 45mm platinum case is a power reserve indicator at 3:00 that is adorned with an 18K white-gold sculpture of Baumgartner himself. Available in a limited edition of ten pieces, this model has a patented gyroscopic module.

ZENITH

ACADEMY GEORGES FAVRE-JACOT – REF. 18.2210.4810/01.C713

Housed in a 45mm 18K rose-gold case, the 797-component El Primero 4810 caliber features a unique horological marvel in a limited edition of 150 pieces. The upper half of the dial reveals the perfectly engineered fusée-and-chain transmission. Ensuring constant force and precision, the complication regulates the energy transmitted from the barrel to the gear train, via a chain between the barrel and fusée, regardless of the power reserve's status. Winding the crown delivers a spectacular treat to the wearer, as a few seconds are enough to watch the chain wrap itself completely around the fusée, ready to reinitiate the 50-hour journey of extraordinary performance. Boasting a silicon escape wheel, the timepiece indicates the small seconds at 7:30 and power reserve at 4:30 within a daringly asymmetrical architecture.

Tourbillons

Girard-Perregaux
Neo-Tourbillon
with Three Bridges

Richard Mille
President and Founder
of Richard Mille

Richard Mille

President and Founder of Richard Mille

"We will continue to enjoy ourselves!"

FOR ALMOST 15 YEARS, THE INDEPENDENT RICHARD MILLE HAS SURPRISED US ALL WITH HIS PLAYFUL WATCHES AND EXTREMELY CUTTING-EDGE TECHNIQUES, IMBUING HIS BRAND WITH HIS PERSONAL PHILOSOPHY. The year 2014 marked not only a triumph in the world of women's watches, but also a new fusion of high athletic performance and horological technique with models such as the RM 36-01 Tourbillon Competition G-Sensor Sébastien Loeb. Though Richard Mille may moderate its productivity in 2015, the brand's inextinguishable dynamism will reserve the right to operate at full blast. Over the course of the year, the brand will unveil numerous new models in several registers: athletic models, lifestyle and women's watches, innovations and high-end pieces.

How would you sum up 2014?
We have experienced unbelievable success! Three years ago, we had sales of 112 million (Swiss francs), 132 million in 2013 and 150 million last year. In 2014, we did a great job of spreading our production and delivery over the course of the year. This year, our goal is 175 million, and next year, 200 million.

How do you explain this success?
First of all, the high end of the market was not affected by the crisis. We also owe our success to a lot of hard work. We covered all the aspects of the business: production, investments in communication, a retail presence as well as research and development. And we are still creative! I always say that economic problems reveal structural weaknesses. Finally, the brand's success stems from work we have done on developing different lines: lifestyle, sports and women's models. I also have other niche markets that I am thinking of exploring.

RM 36-01 Tourbillon Competition G-Sensor Sébastien Loeb

How many watches do you produce each year?

In 2015, we will pass the bar of 3,000 watches. With an increase of about 15% each year, we are not aiming for mass production, but instead controlled growth. Demand continues to exceed supply. Even though the public is hungry for pieces, we wait for the following year to deliver them. On the other hand, the price of our watches is going up. The most high-end models represent our best sales, and they are becoming more and more complicated, and thus more and more expensive.

What does it mean to be an independent watchmaker today?

It's great! The pleasure of working, the rapidity with which we can realize our concepts, the freedom of operation, the control over our growth. We don't feel any external pressure, so we can let our brand develop naturally, with well-considered investments. We've passed all of our development milestones while remaining firmly in the black.

RM 50-01 Lotus F1
Team Romain Grosjean

From an initially small company, Richard Mille has grown quite a bit. Has its philosophy changed?

Today, we are a well-established, international brand. We have a solid foundation to rely on, and we still enjoy a family-style atmosphere. I like to combine my life philosophy with my business philosophy. We're not opportunistic; instead, we develop relationships with wonderful people, and that attitude permeates our offices, our factories, our distribution partners, our contractors, our brand ambassadors, all the way to the end client! From the very beginning, I've said that anyone who comes into our boutiques, even if they don't have the financial means to purchase a watch, should be treated with respect as a watch enthusiast. It means a lot to me that people love our brand.

RM 56-02 Sapphire

Between highly technical pieces and women's watches, which models had the greatest impact in 2014?

It took me years to introduce women's watches. Now they account for 20% of our sales, and I would like to bring that number to 35%. The masculine side of the brand is, however, as strong as ever. Alongside our best-sellers, such as the RO 011, we have launched extremely complex new models. For example, the RM 36-01 Tourbillon Competition G-Sensor was created in collaboration with Sébastien Loeb, who wears it in competition. It integrates the G-Sensor that we originally used in the RM 036 Tourbillon Jean Todt. The mechanism allows pilots to see if the level of Gs is low or dangerously high. In addition, we released the RM 50-01 Tourbillon Chronograph G-Sensor Lotus F1 Team Romain Grosjean. A new version of the RM 56-2 Sapphire has also come out—it was very difficult to create, and our clientele is going crazy for it.

What does Richard Mille have in store for 2015?

Many new models! We have a lot of three-dimensional work, with technical achievements, playful pieces and women's watches. The common denominator is the strong technique in every piece. We will spread out our product launches over the year in order to maximize the media coverage. I don't want to repeat the bad experience we had with the RM 031 High Performance. That particular watch approached the horological Holy Grail by providing a degree of accuracy close to that of quartz technology. We presented it along with several other new models, and the media almost completely ignored it! So this year, you will see our new pieces being unveiled one at a time all year long. This way, we can also time the release of new models to coincide with the deliveries of the products.

Has public taste become less concerned with expectations of precision?

I think that the haute horology market has changed a lot. Our clients want a watch that keeps time, of course, but the most important quality for a piece is that it fits with one's lifestyle, or that it has a playful aspect to it. But I am personally obsessed with performance. We are going to continue to develop these essential concepts: they are the basis for our legitimacy as a watchmaker. Moreover, the products we create with Audemars Piguet Renaud & Papi are fantastic in terms of their contributions to the ongoing history of horology.

What is your distribution strategy?

We have been working right from our beginnings with the same four partners covering four geographical zones and who are now friends. We are very careful in the allocation of our sales and the opening of our boutiques, which are all profitable! At the end of 2014, we had 25 boutiques. Eventually, I project between 40 and 45. We have several possible avenues of development, such as locations in Eastern Europe or New York. In addition, about one hundred retailers offer our watches around the world.

What challenges do you see for Richard Mille in the future?

Maintaining our goal of controlled growth, and not getting carried away. Continuing to perform our fundamental research. And most of all, not denying ourselves anything and continuing to enjoy ourselves!

RM 037 Ladies

RM 07-01 Ladies

Tourbillons

In the language used by francophone astronomers at the end of the 18th century, the word “tourbillon” designated the movement of the stars. That is why Abraham-Louis Breguet used the word for his inspired mechanism, designed to cancel out any fluctuations in the watch’s function due to Earth’s gravitational pull using a small rotating carriage enclosing the regulating organ. Since that time, the tourbillon has become one of the highest expressions of horological savoir-faire, and over the last two decades, it has experienced unprecedented popularity. However, today’s watchmakers are not content with merely reproducing this sophisticated mechanism, originally created for pocket watches. They have contributed to its evolution by developing new technical solutions more befitting wristwatches—double tourbillons, multi-axis tourbillons, new high-tech materials, etc. The fascination of this small carriage, spinning like a heavenly body, also inspires new stagings that experiment with transparency and depth.

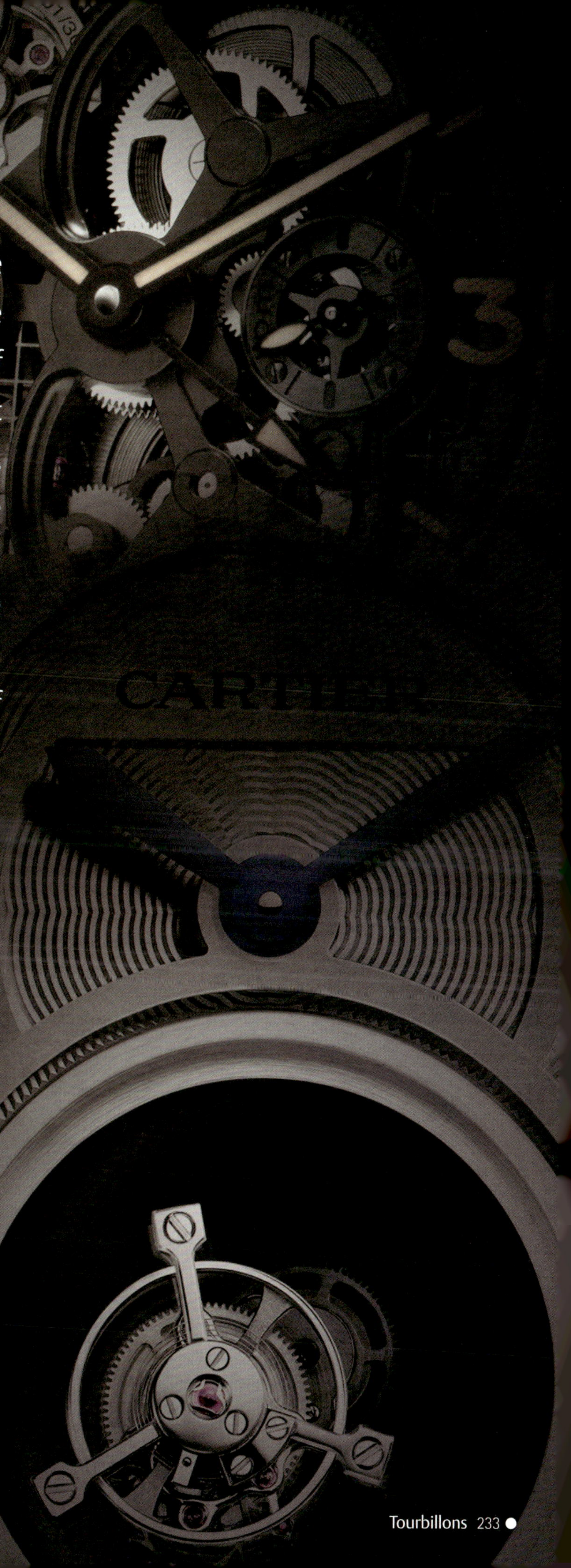

ABRAHAM-LOUIS BREGUET INVENTED THE TOURBILLON to compensate for fluctuations in a pocket watch's functioning due to gravity.

Pocket watches were usually carried in a special pouch or fob on a trouser belt, waistcoat or jacket—hence the name. Thus maintained in a vertical position for many hours, the mechanism—starting with the balance and the balance-spring, two necessary components for accuracy and precision—underwent the detrimental effects of gravity. To avoid such irregularities, the famous watchmaker Abraham-Louis Breguet (1747-1823)—one of the greatest horological geniuses of all time—had the idea of enclosing the "heart" of the watch (balance, balance-spring and escapement) within a small mobile carriage spinning on its axis within a predetermined time (generally one, four or six minutes). It was for this rotation that he chose to name his invention "tourbillon"—a term from astronomy synonymous with "planetary system" among 18th-century scientists. Breguet submitted the patent for his "watch compensating for all the inequalities that occur in the balance and balance-spring" on the 7th of Messidor in the 9th year of the Republican calendar—otherwise known as June 6, 1801. According to the house's archives, 35 tourbillon watches were sold between 1805 and 1823, and throughout the 19th century the tourbillon remained an extremely rare complication, reserved for a few exceptional timepieces.

The 20th-century triumph of the wristwatch could have brought about the disappearance of the tourbillon, as the complication loses much of its significance when the watch moves in all directions with the movement of the wearer's wrist. However, the revival of mechanical horology in the 1980s propelled this mechanism into the firmament of horological complications by positing it as the *nec plus ultra* of Swiss savoir-faire, conferring upon any watchmaker entry into the ranks of the great. Since the turn of the millennium, the tourbillon has become wildly popular. Whereas only 670 tourbillons were produced between 1801 and 1986, today the global production stands at well over 1,000 per year. Over 50 Swiss brands offer them among their collections, which is surprising when one considers that only a small handful of artisans are really capable of making this small magic carriage... Faced with this (relative) banalization, the major watch firms are determined to offer "in-house" tourbillons featuring new technical and aesthetic developments.

HOW DOES IT WORK?

Placed horizontally, a watch's regulating organ (balance, balance-spring and escapement) is uniformly affected by the Earth's gravitational pull. However, when it is placed vertically (and to a lesser extent, in any position besides the horizontal), gravitational force causes certain fluctuations in its functioning and oscillations. These small accelerations and decelerations cause irregularities in the watch's functioning. The idea behind the tourbillon is to place the regulating organ in a small mobile carriage that completes one rotation in a given time, usually one minute. By successively adopting all possible positions, the carriage allows the small irregularities to cancel each other out. The tourbillon carriage is a tiny construction that may contain 70 components with a weight of less than 1g. This is why the production, assembly and setting of a tourbillon watch require exceptional expertise, reserved for the most elite watchmakers and incompatible with mass production.

TOURBILLONS AND PRECISION

The tourbillon was created for the pocket watch, which spends most of its time in a vertical position. It is less necessary on a wristwatch, when the daily movements of the wearer's wrist play the same role as the tourbillon in putting the regulating organ in a variety of positions. Some people go so far as to say that the tourbillon is completely useless in its classic form. To solve this problem, a new generation of tourbillons has arrived—with diverse variations on the manners, axes and speeds of rotation—that supposedly offers enhanced performances for wristwatches. Nonetheless, even though the tourbillon is not in itself synonymous with increased accuracy, the expertise necessary to create such a watch, and the extreme care involved in its production, are both guarantees of a watch that offers superior performance.

FLYING TOURBILLON

Normal tourbillons are supported by two bridges, one above the mechanism and the other below. "Flying" tourbillons are supported by only one, below the mechanism, which allows the watchmaker to give the construction a more ethereal and resolutely virtuoso appearance.

TOURBILLON AND KARUSSEL

Patented by the Danish Bonniksen in 1892, the karussel (or "carrousel") has the same ultimate goal as the tourbillon. However, it is distinguished (or could be distinguished—the experts do not all agree) by the double wheel train that links it to the barrel: the first provides the necessary energy for the functioning of the escapement, and the second controls the speed of the carriage's rotation.

A Reinvented Classic

With its mobile carriage spinning on its axis like a planet, the tourbillon provides a fascinating sight that most brands are keen to exhibit through the dial, traditionally through an aperture at 6 o'clock. But the ultra-sophisticated mechanism is now the object of numerous technical and aesthetic innovations, such as the use of new materials that offer enhanced lightness, or the search for unusual dial configurations.

Breguet
Classique Tourbillon Extra-Thin Automatic 5377
A worthy heir to Abraham-Louis Breguet's famous patent, the Classique Tourbillon Extra-Thin Automatic 5377 combines a silicon balance-spring and lever-wheel, a titanium balance-wheel and tourbillon carriage, and a platinum peripheral rotor—all housed within an extraordinarily slim (3mm) movement.

Parmigiani, *Pershing Tourbillon Abyss*

A combination of titanium and pink gold, the Pershing Tourbillon Abyss by Parmigiani displays its 30-second tourbillon on a dark blue dial adorned with a wave motif—a nod to the famous Italian yacht maker after which the collection is named.

Louis Moinet, *Derrick Tourbillon*

Louis Moinet's Derrick Tourbillon features a double animation on the dial: the tourbillon rotating at 6 o'clock is joined by an automaton that depicts a derrick drilling an oil well.

Antoine Preziuso, *GTS Grand Tourbillon Sport*

Equipped with a cobalt chrome case, whose aerodynamic forms evoke the sportscars that inspired them, Antoine Preziuso's GTS Grand Tourbillon Sport is also notable for its large red hands, which recall the needles on speedometers.

H. Moser & Cie, *Venturer Tourbillon Dual Time*

The first tourbillon watch from H. Moser & Cie, the Venturer Tourbillon Dual Time houses an automatic movement made in house with a double Straumann® balance spring to improve its precision and an interchangeable tourbillon module to facilitate maintenance. The refined dial also displays a second time zone.

A. Lange & Söhne, *Richard Lange Tourbillon "Pour le Mérite*

Beneath the regulator-type dial of the Richard Lange Tourbillon "Pour le Mérite" watch lie two mechanisms initially designed to improve timekeeping precision: a tourbillon and a fusée-and-chain energy transmission system.

Patek Philippe, *10-Day Tourbillon, Ref. 5101*

Endowed with a power reserve of ten days, the 10-Day Tourbillon, Ref. 5101 from Patek Philippe is notable for its cambered Art Deco-style case. As with all its tourbillon watches, the manufacture prefers not to show the tourbillon from the dial side, for fear that sunlight might dry up the lubricating oils.

Graff, *MasterGraff Ultra Flat Tourbillon*

The MasterGraff Ultra Flat Tourbillon by Graff features a 7.05mm-slim case housing a 3.5mm-thin flying tourbillon movement. The bezel is distinguished by the brand's signature diamond-facet design.

Montblanc, *ExoTourbillon Chronographe*

On its ExoTourbillon Chronographe, from the Villeret 1858 collection, Montblanc has placed the balance on the exterior of the tourbillon carriage. As the balance is larger, its amplitudes gain more precision, and the tourbillon carriage, freed from the weight of the balance, requires less energy for its rotation.

IWC

Ingenieur Constant-Force Tourbillon

As its name implies, the Ingenieur Constant-Force Tourbillon by IWC has a constant-force mechanism guaranteeing virtually invariable amplitude of the balance and thus enhanced precision.

Room for Transparency

Certain brands have chosen to distill the tourbillon's magic with constructions whose heart is open, revealing the entire interior of the movement—masterpieces of skeletonization that push the frontiers of possibilities, or fascinating transparent stage-settings within which the mobile carriage seems to float weightlessly.

Richard Mille, *RM 27-01 Rafael Nadal*

Designed in cooperation with the international tennis star, the RM 27-01 Rafael Nadal by Richard Mille combines a feather-light weight (3.5g for the movement) with a high-tech construction based on a carbon case and a cable-suspension mechanism.

Blancpain, *Le Brassus Carrousel Minute Repeater Flyback Chronograph*

Blancpain has made an innovative move by rehabilitating the carrousel (also spelled karussel), a mechanism akin to the tourbillon (see "How does it work?" section), and incorporating it into several wristwatches, including this Le Brassus model, which boasts a one-minute flying carrousel at 6 o'clock, minute repeater and automatic-winding flyback chronograph.

De Bethune, *DB 28T*

The DB 28T from De Bethune stands out for its futuristic architecture and its ultralight tourbillon in titanium and silicon. The tourbillon carriage rotates over 30 seconds and the balance oscillates at 36,000 vph.

Vacheron Constantin
Patrimony Traditionnelle 14-Day Tourbillon Openworked

Thanks to its four barrels stacked in pairs, the Patrimony Traditionnelle 14-Day Tourbillon Openworked boasts a power reserve of 336 hours. The movement is entirely openworked according to a design that emphasizes the effect of depth in the watch.

Manufacture Royale, *Androgyne*

The brand Manufacture Royale (named for the watch company founded by Voltaire in 1770) features in its Androgyne line its own manual-winding tourbillon movement equipped with a silicon lever and escape-wheel.

Corum

Golden Bridge Tourbillon

Corum shows off its famous "baguette" movement in this Golden Bridge Tourbillon model, with a sapphire bridge. In order to reveal the flying tourbillon, a part of the movement (usually in gold) has been replaced by a sapphire component.

Girard-Perregaux, *Neo-Tourbillon with Three Bridges*

In 1889, Girard-Perregaux made its mark on watchmaking history by creating its famous "tourbillon with three gold bridges." Through its Neo-Tourbillon with Three Bridges, the watchmaker revisits this heritage, combining entirely redesigned titanium bridges with a uniquely shaped sapphire crystal and a new movement architecture.

Bovet 1822

Amadeo® Fleurier Tourbillon Virtuoso III

The Amadeo® Fleurier Tourbillon Virtuoso III from Bovet 1822 associates a tourbillon and a retrograde perpetual calendar within a completely transparent architecture. The ingenious case design allows one to easily transform this re versible wristwatch into a table clock or pocket watch.

Audemars Piguet

Jules Audemars Large Date Tourbillon

The Jules Audemars Large Date Tourbillon from Audemars Piguet, powered by a manual-winding in-house caliber, opts for understated elegance with a large tourbillon carriage and a panoramic date display at 12 o'clock.

Revelation

R04 Tourbillon Magical Watch Dial®

The R04 Tourbillon Magical Watch Dial® by Revelation comprises a patented system serving to make the movement appear as if by magic through a dial that has turned transparent. It is also equipped with a Tourbillon Manège® caliber featuring a rotating regulating system in which the regulating organ is mounted on a mobile arm sandwiched between two wheels mounted on the central axis.

The Third Dimension

Recent years have also witnessed the appearance of a new generation of tourbillons equipped with inclined carriages or several simultaneous rotational axes, and/or revolving at speeds that diverge from the traditional one minute (often 30 or 24 seconds). Besides their often spectacular appearance, these three-dimensional constructions are designed to be more suited to wristwatches, by multiplying the positions of the balance-spring with the case.

Harry Winston, *Histoire de Tourbillon 4*

The Histoire de Tourbillon 4 model by Harry Winston is equipped with a tri-axial flying tourbillon starring beneath a domed sapphire crystal, featuring an inner carriage rotating in 45 seconds, an intermediate carriage in 75 seconds and an outer carriage in 5 minutes.

Vianney Halter, *Deep Space Tourbillon*

Vianney Halter has equipped its Deep Space Tourbillon with a tourbillon that rotates on three axes at once—all within a design that evokes space exploration, with a titanium case and dome-shaped sapphire crystal. The hour and minute are indicated by two curved hands in blued steel that extend inward from the periphery of the dial.

Concord, *Quantum C1 Gravity*

Concord created shock waves with its launch of the Quantum C1 Gravity, equipped with an unprecedented bi-axial tourbillon positioned vertically, i.e. almost outside the movement, and linked to the bottom plate by a series of cables reminiscent of suspension bridges.

Greubel Forsey, *Art Piece 1*

On the Art Piece 1 by Greubel Forsey, the Double Tourbillon 30° (with its inner carriage rotating in one minute and the outer carriage in four minutes), vies for attention with a microsculpture created by contemporary artist Willard Wigan.

Panerai, *Radiomir Tourbillon GMT Ceramica Lo Scienzato*

On the Radiomir Tourbillon GMT Ceramica Lo Scienzato from Panerai, the carriage turns at a right angle to the oscillation axis of the balance, performing two turns per minute. Its in-house Caliber P.2005/S is also distinguished by a substantial six-day power reserve, achieved thanks to the energy from three barrels.

Jaeger-LeCoultre, *Duomètre Sphérotourbillon*

Equipped with a Dual-Wing movement, Jaeger-LeCoultre's Duomètre Sphérotourbillon can rightfully claim to be the first tourbillon watch adjustable to the second. The "sphérotourbillon" rotates on two axes: one rotation is completed in 30 seconds, and the other (on a 20° angle) in 15 seconds.

Thomas Prescher
Mysterious Automatic Double Axis Tourbillon

On the Mysterious Automatic Double Axis Tourbillon from Thomas Prescher, only the tourbillon, the calendar, the oscillating weight and the drums serving as displays are visible. The movement is completely hidden by the bezel, in the left and right sides of the case. Placed on a rotation axis by means of a dedicated arm, the centered tourbillon turns on two axes.

Double Revolutions

A few innovative models offer a sophisticated system in which the tourbillon completes a rotation in one minute, while performing a complete turn of the dial in one hour (or more)—mirroring the double rotation of the Earth on its axis and around the sun. These are truly magical merry-go-rounds.

Cartier
Rotonde de Cartier Astrotourbillon Carbon Crystal

The Rotonde de Cartier Astrotourbillon Carbon Crystal has an original construction with a one-minute central tourbillon fitted above the movement. The arrow-shaped balance bridge indicates the seconds. The case is composed of a niobium-titanium alloy, while the tourbillon bridges, lever and the escape-wheel are in carbon crystal.

Ulysse Nardin, *Freak Phantom*

The Freak Phantom from Ulysse Nardin stands out for its very innovative construction. The movement rotating within the case allows the watch to display the hours and minutes with neither hands nor dial. The flying karussel tourbillon indicates the seconds. The escapement and balance spring are in silicon.

Bernhard Lederer, *Gagarin Tourbillon*

A tribute to the famous Soviet cosmonaut, the Gagarin Tourbillon by Bernhard Lederer displays an orbital 60-second tourbillon also rotating counter-clockwise in 108 minutes. A prominent loupe, held in place by a catch inspired by a hatch on the Vostok space capsule, affords a chance to admire the dial details.

Piaget, *Polo Tourbillon Relatif Limelight Jazz*

The principle of a double revolution applies to Piaget's Polo Tourbillon Relatif Limelight Jazz. Suspended from the tip of the minute hand, the mobile flying tourbillon carriage appears to be disconnected from the movement driving it, creating a fine encounter between mechanical prowess and visual magic.

Cartier, *Rotonde de Cartier Mysterious Double Tourbillon*

The Rotonde de Cartier Mysterious Double Tourbillon is distinguished by its flying tourbillon featuring a double rotation (once on its own axis in 60 seconds, and around an aperture every 5 minutes) and of which the carriage appears to be floating in mid-air thanks to a transparent disc-based system.

Watches with Several Tourbillons

Another technical and aesthetic path explored by some watchmakers is the multiplication of tourbillons within a single watch. Horologers come up with double, even quadruple tourbillons, tied together by differential gear systems that average out the functioning of different regulating organs.

Breguet, *Twin Rotating Tourbillons*

The Twin Rotating Tourbillons from Breguet boasts two independent tourbillons coupled by a differential gear and fixed to a bottom plate completing a full rotation in 12 hours. These two mechanisms mutually compensate for their infinitely small variations in rate in order to offer superior precision.

Greubel Forsey, *Invention Piece 2*

A reinterpretation of its 2008 Quadruple tourbillon à différentiel, Greubel Forsey's Invention Piece 2 possesses two double tourbillon carriages (a first carriage executing a complete rotation in one minute, contained within a second carriage effecting a complete revolution in four minutes), linked by a spherical differential placed at the center of the timepiece.

TAG Heuer, *Carrera Mikrotourbillons*

The Carrera Mikrotourbillons from TAG Heuer possesses two tourbillons, one for the time, the other for the 1/100-of-a-second chronograph. The latter, described as "the fastest tourbillon in the world," beats at 360,000 vph and completes 12 revolutions per minute.

Roger Dubuis
Hommage Double Flying Tourbillon

Equipped with two rotating carriages linked by a differential gear, the Hommage Double Flying Tourbillon by Roger Dubuis possesses an elegant hand-guilloché dial. The 452-component movement is honored by the prestigious Poinçon de Genève.

Jacob & Co., *Napoleon Quadra Tourbillon*

Crafted in titanium, the imposing Napoleon Quadra Tourbillon from Jacob & Co. features no fewer than four skeletonized tourbillons and four complete time zones with hour and minute displays.

Corum, *Admiral's Cup AC-One 45 Double Tourbillon*

The Admiral's Cup AC-One 45 Double Tourbillon from Corum displays its two mobile carriages in two apertures at the bottom of the dial, joined by a retrograde date at the top of the dial and an ingenious system of correcting the time display via a pusher in the crown.

The Feminine Tourbillon

Within just a few years, women's tourbillon watches have accomplished an astonishing breakthrough. Uniting haute horology and—most often—haute jewelry, the most prominent brands have undertaken to create timepieces that set a refined stage for the little "magic carriage."

> **Parmigiani**, *Kalpa XL Tourbillon Diamants*

The Kalpa XL Tourbillon Diamants offers a one-week power reserve and a carriage performing two rotations per minute. The tonneau-shaped case, inspired by the golden ratio, is adorned with a precious diamond pavé setting.

Ulysse Nardin, *Royal Ruby Tourbillon*

Adorned in rubies and diamonds, the Royal Ruby Tourbillon by Ulysse Nardin sets a fine stage for its flying tourbillon within a supremely transparent structure reflecting the red glow of the precious stones. This platinum limited edition is available in 99 pieces.

Chanel, *Première Flying Tourbillon*

An alliance of aesthetic refinement and technical complexity, Chanel's Première Flying Tourbillon associates an octagonal dial (evoking the stopper for the brand's N° 5 perfume) with a tourbillon carriage shaped like a camellia, emblem of the Parisian house.

Louis Vuitton
Tambour Monogram Tourbillon

The Tambour Monogram Tourbillon by Louis Vuitton plays on the House signature codes, including the famous LV initials. The tourbillon carriage, visible at 6 o'clock, is equipped with a bridge shaped like the flower on the famous Monogram canvas.

CHRISTOPHE CLARET

X-TREM-1 – REF. MTR.FLY11.090-098

The X-TREM-1 is as technologically advanced as it is visually outstanding. First released in 2012, this model presents a revolutionary way to tell time. The hours and minutes are displayed in two hollow steel spheres isolated in sapphire tubes on both sides of the caseband. The spheres appear to be floating and are controlled by magnetic fields generated by two magnets moved by very thin pieces of surgical silk thread. This model is a new addition to the X-TREM collection and is complete with a flying tourbillon which is angled at 30° on a convex titanium mainplate. The timepiece is available in a limited edition of eight pieces.

CORUM

ADMIRAL'S CUP AC-ONE 45 DOUBLE TOURBILLON – REF. A108.02337

The Admiral's Cup now introduces a brand new movement: the manual-wound Caliber CO 1008. This mechanism features two parallel-mounted tourbillons combined with a retrograde date and a time-adjustment system. The generous 72-hour power reserve is dedicated to ensuring remarkable reliability. By simply pressing the crown pushbutton, the wearer can bring the minutes hand to the 12th hour marker to set the exact time; this can be done up to five minutes before or after a full hour. The grade-5 titanium case of the Admiral's Cup AC-One 45 Double Tourbillon stylishly displays a 45mm diameter enhanced by the immediately recognizable signature Corum design. This piece is water resistant to 30atm.

DE BETHUNE

DB28ST BLACK – REF. DB28STTS5PN

Available in a 20-piece limited edition, De Bethune's DB28ST Black exhibits all the technical innovations that have earned the brand the admiration of all watch connoisseurs: the self-regulating twin barrel, the silicon/white-gold balance wheel and balance-spring with flat terminal curve, the central jumping seconds, the silicon escape-wheel and the De Bethune silicon/titanium 36,000vph 30-second tourbillon. With a space-age design, the DB28ST combines a 43mm mirror-polished grade-5 titanium case with a platinum bezel. The black dial flaunts its own contradictions, displaying both Roman and Arabic numerals on a grained and engraved sterling silver chapter ring as well as a 30-second tourbillon and power reserve indication at 6:00. The DB2119 caliber provides four days of power reserve.

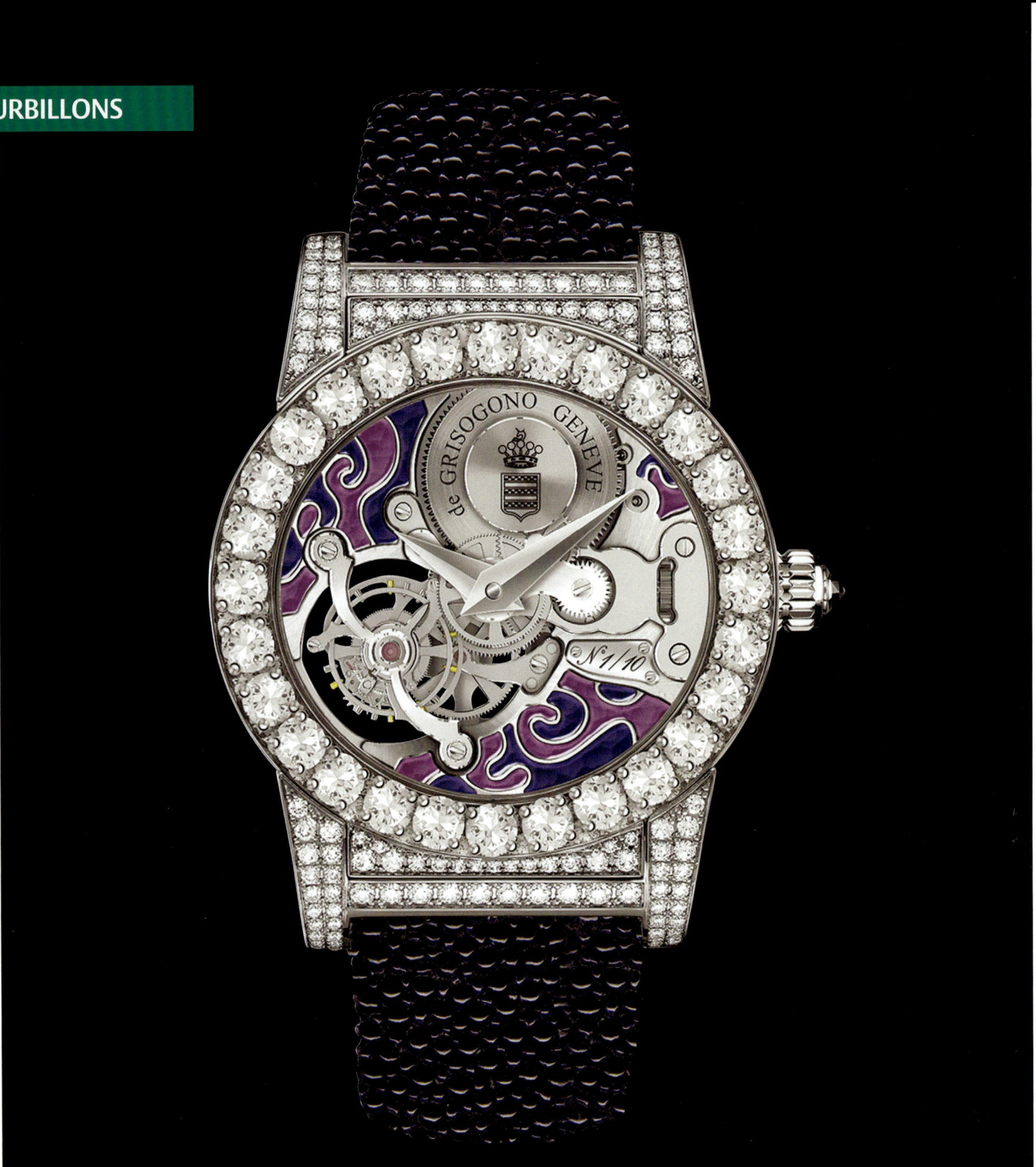

de GRISOGONO

TONDO TOURBILLON GIOIELLO – REF. S05

The diamond-swathed Tondo Tourbillon Gioiello houses an exclusive new in-house developed and assembled de Grisogono mechanical movement, fitted with an off-centered tourbillon at 8:00. The exclusive DG 31-88 caliber specially designed by de Grisogono makes a perfect fit with the oval-shaped Tondo XL case, and is exquisitely enhanced with various finely crafted technical details. The hand-polished and beveled structural components reflect the noble traditions of fine horological finishes. The brand signature and its coat-of-arms are engraved on the barrel ratchet-wheel at 12:00, and the various parts of the main plate serve as the dial. The bridges, tourbillon carriage and barrel, all developed in-house, are also clearly visible. This watch with 72-hour power reserve displays the hours and minutes.

de GRISOGONO

TONDO TOURBILLON GIOIELLO – REF. S06

This model introduces the first ever in-house mechanical hand-wound movement equipped with a tourbillon, unusually positioned at 8:00. The white-gold case is water resistant to 30m and sparkles with more than 8 carats of diamonds, running along the sides and all the way up to the lugs. The so-called "four grain" gem-setting technique keeps the quantity of metal beneath the gems to a minimum, thereby intensify the sparkling reflections. 33 larger diamonds frame the bezel and endow it with remarkable radiance, while the brand's emblematic deep black diamond adorns the crown. The daring architecture of the Tondo Tourbillon Gioiello, as well as its exclusive caliber, embodies de Grisogono's vibrant creativity.

F.P. JOURNE

TOURBILLON SOUVERAIN

The exceptional 18K rose-gold mechanism of the Tourbillon Souverain, visible through the sapphire crystal, is the only one that uses the remontoire or constant-force device system for its timekeeping performances. This complication, representing a unique feature on a contemporary wristwatch, provides a more accurate read-off of time. It also features an independent seconds system called "seconde morte" or "dead-beat seconds." This name stems from the fact that the hand remains motionless ("dead") as long as the second has not actually elapsed. The case in platinum houses a white-gold and silver dial and an 18K rose-gold movement.

GIRARD-PERREGAUX

BI-AXIAL TOURBILLON – REF. 99810-81-000-BA6A

For this model, the Tourbillon Bi-Axial has chosen tantalum, a durable, malleable and corrosion-resistant gray-blue metal, which is usually used in the chemical, electronic and aerospace industries. In addition to its many technological assets, it also lends a masculine and contemporary style to the watch. The pure design and strong character of the timepiece are conveyed by the style of the three bridges cut in sapphire. The bi-axial tourbillon mechanism boasts surprising delicacy: the two cages, featuring more than 110 parts, only weigh 0.8g. The platinum is decorated by a traditional circular guilloché motif, counterbalanced by its galvanic gray finishing, while the barrel bears the characteristic elliptic pattern of the famous three-parallel-bridge pocket watch made by Constant Girard in the 1860s. Available in a limited edition of 12 pieces.

GIRARD-PERREGAUX

CAT'S EYE TOURBILLON WITH GOLD BRIDGE – REF. 99490D53P703-CKHA

The feminine curves of the Cat's Eye Tourbillon with Gold Bridge sparkle with 308 beautifully set diamonds from the bezel to the lugs and even on the crown. Its distinctive oval case in white gold frames a superb white mother-of-pearl dial. Gently rounded leaf-shaped hands float elegantly above a floral motif engraved in relief on the dial. At 6:00, the iconic Girard-Perregaux Tourbillon reigns majestically under a Gold Bridge delicately carved with a floral motif that echoes the dial. It precisely regulates the running of the manually wound calibre GP09700-0012. Containing 174 parts decorated, adjusted and assembled by hand, its perfect finishing is another sign of the exacting standards of Girard-Perregaux, a pioneer in producing beautiful mechanical watches for women.

GIRARD-PERREGAUX

GIRARD-PERREGAUX TOURBILLON WITH THREE GOLD BRIDGES, LADY REF. 99240D-52-A701-CK7A

The renowned Girard-Perregaux Tourbillon with Three Gold Bridges enjoys a new interpretation in this feminine version. Crafted in pink gold, the subtly profiled case gracefully fits the most delicate wrists. The bezel and flange are set with more than 1.8 carats of diamonds, illuminating the tourbillon and its bridges. Their perfect geometry is underscored by entirely hand-polished finishing, while the gentle, rounded proportions seem to be magnified by an exceptional light-reflecting mirror polish. This 38mm ladies' version sports a delicately cut mother-of-pearl dial positioned on the three bridges. The blue alligator strap is complete with a buckle set with 18 diamonds.

GIRARD-PERREGAUX

NEO-TOURBILLON WITH THREE BRIDGES – REF. 99270-52-000-BA6A

The Neo-Tourbillon with Three Bridges watch features a completely redesigned caliber that is larger than previous versions, yet respects the identity codes of the Tourbillon with Three Bridges: it features a symmetrical design, an aligned gear-train, a tourbillon case at 6:00 and a micro-rotor mounted under the barrel at 12:00. In addition to these features, the GP09400 caliber (consisting of 245 components) incorporates several exclusive developments. Its frequency is of 3Hz to ensure excellent performance and longevity. It has also been designed in a modular fashion to optimize its assembly and maintenance.

GIRARD-PERREGAUX

TOURBILLON WITH THREE GOLD BRIDGES – REF. 99193-52-002-BA6A

Born in the hands of Constant Girard-Perregaux in the 19th century, the Three Bridges movement is a revolution in the history of fine watchmaking. With it, the mechanism becomes a design feature in its own right, extending beyond a solely technical function. Within a 41mm diameter, the three, arrow-shaped, parallel bridges form the perfect balance. Their geometry allows for no mistakes as any imperfection would be revealed. At 12:00, the barrel drum is meticulously skeleton-crafted to reveal the oscillation of the elegant and ingenious platinum micro-rotor. The tourbillon system is also the subject of particular care as its 80 components need to be assembled in a 10mm diameter. The cage, escapement and pallets bridge are chamfered and hand-polished, as are the gold adjustment screws.

GIRARD-PERREGAUX

TRI-AXIAL TOURBILLON – REF. 99815-52-251-BA6A

This exclusive model, available in a limited edition of just ten pieces, signifies a major feat for Girard-Perregaux in its quest for ultimate precision with its high-speed tourbillon on three separate axes. Tourbillons have been a specialty at Girard-Perregaux for over 150 years, but this model truly demonstrates the Manufacture's mastery. At the heart of the timepiece is a traditional tourbillon. Its cage performs one rotation per minute and is integrated within a structure positioned on a second axis, which rotates once every 30 seconds. Both are then inserted into a third system which rotates every two minutes. The Tri-Axial Tourbillon also displays an off-centered hours and minutes counter and a power reserve indicator.

GUY ELLIA

TOURBILLON MAGISTERE

Designer Guy Ellia's famously skeletonized mysterious-winding tourbillon is now available in rose gold. The Time Square Tourbillon Magistère's Caliber TGE 97 was created by Swiss manufacture Christophe Claret and features a 110-hour power reserve, one-minute tourbillon, a bottom plate and bridges. The movement beats at 21,600 vph within an entirely hand-chamfered cage.

GUY ELLIA

TOURBILLON MAGISTERE II

Guy Ellia has renewed its collaboration with Christophe Claret to present their timepiece enhanced with a tourbillon escapement: the Tourbillon Magistère II. Inside its distinctive case, an elongated rectangle (44.2x36.7mm), the skeleton movement features mysterious winding and features an "invisible" dial, whose main hands display the hours and minutes while seeming to float in space. In fact, the technical genius of this watch, produced in a limited edition of 12 pieces for each version, lies in the inversion of its bridges compared to a classic tourbillon timepiece. Hence the location of the winding crown at 6:00 and the use of the crown positioned at 12:00, while maintaining pressure on the security pushbutton at 10:00 to set the time. This piece is for lovers of fine watchmaking in search of the unusual.

GUY ELLIA

TOURBILLON ZEPHYR

With a movement created by Swiss manufacture Christophe Claret, the Tourbillon Zephyr symbolizes transparency with excellence. Caliber GES 97 features a 110-hour power reserve and winding ring set with 36 baguette-cut diamonds (1.04 carats), or engine turning and one-minute tourbillon movement. The latter beats at 21,600 vph within an entirely hand-chamfered cage. The convex case with 950 platinum sides has a transparent back cover and hearth engraving. Mounted on an alligator strap with a folding buckle, this model is also available in platinum and titanium.

HUBLOT

CLASSIC FUSION TOURBILLON FIRMAMENT OSMIUM – REF. 502.CX.0002.LR

In a world-first, Hublot has crafted a watch with a dial made from osmium crystal. As well as being the rarest metal on Earth, osmium is also very hard; it is also the densest and heaviest metal. Osmium crystals vary in size from tenths of a millimeter to just a few millimeters, making osmium in its crystal form one of the greatest new developments in the precious metals market. Osmium is crystallized using a highly sophisticated process to heat it to over 3,000 °C. The result is breathtakingly unique and its natural brilliance is as intense as the heavens, hence the name of the watch—Classic Fusion Tourbillon Firmament. Powered by the HUB6017 caliber, this piece has a power reserve of five days.

HUBLOT

CLASSIC FUSION TOURBILLON NIGHT OUT – REF. 505.CS.1270.VR

Available in a limited edition of 30 pieces, Hublot's Classic Fusion Tourbillon Night Out takes its style cues from the always-fashionable tuxedo. 45mm in diameter, made from black polished ceramic, with a black lacquered dial set with 10 baguette diamonds for an exceptional luster, two polished, rhodium-plated hands and an opening at 6:00, this timepiece exudes chic simplicity. Exposing the magic of a flying tourbillon on its dial, it also possesses a sapphire caseback through which the HUB6010 movement, which boasts 155 components and a five-day power reserve, can be seen. The movement is entirely developed and produced in-house by the engineers and watchmakers of the Hublot manufacture. The watch sits on a strap of shiny black calfskin sewn onto rubber, for comfort and flexibility, making it the perfect accessory for the gentleman in evening wear.

HUBLOT

CLASSIC FUSION TOURBILLON SKULL – REF. 505.UC.0170.VR SKULL

The ceramic-coated aluminum of the Classic Fusion Tourbillon Skull gives this piece a truly striking appearance. This material, which is new to Hublot, has excellent resistance to corrosion and friction, and is twice as light as ceramic. Hublot's famous manual-winding skeleton tourbillon movement, HUB6014, with its five-day power reserve, has also seen several transformations. This is particularly true of its bridges and plate, which have undergone 3-D machining: a laser machining phase has transformed the flat bridges, creating a rare three-dimensional relief on these components, which are usually flat. One last small detail adds further interest to the piece: the hour markers are Roman numerals, which is also a first for Hublot. This unconventional piece is available in a limited edition of 50 pieces.

HUBLOT

CLASSIC FUSION TOURBILLON VITRAIL – REF. 502.NX.0003.LR

The inspiration behind this watch was simple: a stained glass window is essentially an outline filled with colored glass. The same creative concept is at play here: the skeleton watch movement serves as a miniaturized, colorful stained-glass window. It features a HUB6017 movement, a skeleton tourbillon comprising 251 components, a five-day power reserve and a contemporary graphic design. Within the construction of the watch, there are no fewer than 15 tiny openings housing a sliver of glass that is secured with O-rings. This is a very high-tech glass that is used in physics to filter the different frequencies or colors of light and to hold its color over time. The first four references in this collection are crafted in titanium or ceramic, with red or blue tones. Each version is available in a limited edition of 20 pieces.

IWC

PORTUGUESE TOURBILLON HAND-WOUND – REF. IW546301

Revealed at 9:00 through an opening in the slate-colored dial, the 98900 caliber's 64-component flying tourbillon brings a taste of the 28,800 vph movement's micro-mechanical intricacies to the understated watch. In addition, the organ's rich golden tones perfectly contrast the sheen of the 43mm case's 18K white gold. The hand-wound timepiece, complete with small seconds at 6:00, reveals its movement's finely decorated nickel-silver three-quarter bridge through a sapphire caseback and is worn on a dark brown Santoni alligator leather strap.

JACOB & CO.

BRILLIANT ARLEQUINO FLYING TOURBILLON – REF. 210.543.40.HX.HX.1HX

The Brilliant Arlequino Flying Tourbillon is a true masterpiece. The finishing of this timepiece, comprised of 206 components with a flying tourbillon and power reserve of 100 hours which can be seen via the power reserve indication on the caseback is a delight for any watch enthusiast. The eye catching sensation of the "arlequino" pattern is the result of the collaboration of multiple gem-cutters and setters. Each piece is unique in the pattern of gem setting which demands an expert eye to select the perfect stones and the highest caliber of gem setting skills and talent to craft the masterful presentation of diamonds and multi-colored sapphires set into the18K rose-gold case.

JACOB & CO.

BRILLIANT SKELETON TOURBILLON ART DECO ICY BLUE – REF. 210.545.30.BC.RB.1BC

The Brilliant Skeleton Tourbillon Art Deco Icy Blue crafts its skeletonized movement through sapphire crystal plating and bridges. All non-essential metal on the mechanical components of this sparkling wristwatch have been trimmed away leaving only the dazzling art deco style bridges set with 178 brilliant cut pink diamonds visible in the openworked movement. Further adding to the brilliance of this piece each icy blue sapphire has been carefully selected to match in color and form before being invisibly set in the 18k white-gold case. The composition of this timepiece is a masterpiece of world class gem setting artistry and spectacular high horology.

JAQUET DROZ

GRANDE SECONDE TOURBILLON AVENTURINE – REF. J013014270

The use of aventurine in dials is nothing new for Jaquet Droz, as the sparkly mineral can be found in many of their pieces. The Grande Seconde Tourbillon Aventurine, however, is the first to associate starry blue aventurine and a tourbillon. The center of the dial is white mother-of-pearl, framed by 18K white gold set with 90 diamonds (0.14 carats). The frame matches the 39mm case, which is also white gold and set with 260 diamonds (1.47 carats). The center of the dial features an off-centered hours and minutes counter, as well as a tourbillon at 12:00. The tourbillon is powered by the Jaquet Droz 25JD caliber.

PARMIGIANI

OVALE TOURBILLON – REF. PFH150-1001400-HA1441

The Ovale Tourbillon combines an ellipse with the famous round cage and its hypnotic oscillation. The PF501 movement, a jewel of Parmigiani Fleurier expertise, is one of the few 30-second tourbillons on the market, rotating twice every minute for increased accuracy. The essence of elegance, the Ovale Tourbillon is available in rose gold with either a black onyx dial or a white-gold dial featuring silvered Côtes de Genève decoration. This exceptional model is available in a limited edition of 60 pieces, featuring 30 of each version.

PARMIGIANI

PERSHING TOURBILLON 30 SECOND AJOURE – REF. PFH552-1010301-HA1442

The new Pershing Tourbillon 30 Second Ajouré model represents the partnership between Parmigiani and Ferretti Group, the prestigious Italian yacht manufacturer. This model exemplifies technical mastery, as the tourbillon cage, visible on both sides of the movement, rotates once every 30 seconds, providing a great deal of setting constancy. The openworked movement makes this model rare in design as the main plate reveals the mechanism wheels and double barrel, with added visibility from the numerous openings on the bridges. This model is available in a limited edition of 40 pieces.

PERRELET

TURBILLON – REF. A3037.1

The name Turbillon fuses the Turbine, one of Perrelet's most sophisticated collections with the Tourbillon, one of haute horology's most prestigious complications. The turbine's distinctive transparent wheel that spins freely above the dial offers unobstructed views of the regulating organ within. The Turbillon introduces a refined turbine constructed of noble sapphire and transparent blades that enliven the dial. This piece is powered by the manufacture made automatic-winding P-371 caliber with an exclusive personalized Perrelet oscillating weight. The double rotor and tourbillon carriage at 12:00 are also accompanied by central hour and minute hands. This model is water resistant to 5atm.

PIAGET

PIAGET EMPERADOR COUSSIN XL TOURBILLON – REF. G0A37039

The world's thinnest automatic-winding tourbillon owes its title to the house's 1270P caliber. Measuring a mere 10.4mm in thickness, this new interpretation of Piaget's aesthetic codes possesses an architecture as spectacular as the exposed movement that powers the watch's heartbeat. Through a laser- engraved sapphire crystal, the wearer may admire an offset white-gold micro-rotor lined with 107 brilliant-cut diamonds (0.2 carat) as well as an imposing tourbillon cage, offering a veritable third dimension to a dial that features a delicate laser-guilloché craftsmanship and beautifully tapered hour markers. The case on this piece makes the model not only remarkable for watch aficionados, but for haute jewelry enthusiasts as well. The 18K white-gold case is set with over 1,000 diamonds weighing in at approximately 16.5 carats.

RICHARD MILLE

RM 19-02 FLYING TOURBILLON FLEUR

This bold piece exemplifies Richard Mille's modern take on timekeeping and also his mastery of both haute horology and haute jewelry. The RM 19-02 contains a magnolia as its leading concept. The beautiful flower evolved millions of years ago and it is fragile in appearance but resilient in nature. The magnolia is seen on the lower left-hand corner of the dial and it surrounds the flying tourbillon escapement with five delicate and colored petals. The magnolia opens and closes via the pushbutton at 9:00 and the flying tourbillon moves upwards 1mm when fully opened, replicating the way the flower arches in nature to increase its chances of pollination. The RM 19-02 is available in 18K white gold with stone setting and is available in a limited edition of 30 pieces.

RICHARD MILLE

RM 51-02 TOURBILLON DIAMOND TWISTER

This high jewelry tourbillon pays homage to the recurring pattern of twisting spirals in nature. We see spirals in beach waves, seashells, plants and the Milky Way and now we also see them on this stunning and intricate timepiece. There are 14 rays of diamonds that emerge from the tourbillon escapement at 6:00. Each ray is perfectly set with a row of single diamonds while the case is fully set with white diamonds. The baseplate of this hand-wound tourbillon is made of onyx and the free sprung balance and skeletonized bridges are made of PVD-treaded grade-5 titanium with black polish. The torque-limiting crown prevents the wearer from accidentally overwinding the piece.

ROGER DUBUIS

EXCALIBUR SKELETON DOUBLE FLYING TOURBILLON – REF. RDDBEX0395

The iconic case of the Excalibur 45 collection demonstrates the appeal of the Excalibur identity codes. At the same time, it provides a new presentation for the skeletonized double tourbillon within which is of outstanding quality and original design. By reducing the plates and bridges to their absolute minimum, Roger Dubuis has created a spacious yet highly technical mechanism, the RD01SQ movement. The Geneva watchmaker thus presents a design with an open heart and a top-flight movement, whose thinner case gives the signature codes a new dynamism.

ROGER DUBUIS

HOMMAGE DOUBLE FLYING TOURBILLON – REF. RDDBHO0563

Endowed with the new RD100 movement, this piece is composed of 452 individually hand-finished parts and a power reserve of 50 hours. It took 1,200 hours to make this piece, which is a testament to the rigorous standards of Manufacture Roger Dubuis. The two flying tourbillons at 5:00 and 7:00 pierce an exquisitely decorated dial, which is complete with hand-crafted guilloché. The pink-gold timepiece which measures 45mm in diameter and 3atm in water resistance combines a sophistication of the double tourbillon with the classic delicacy of artisanal guilloché craftsmanship on an extremely eye-catching model.

TAG HEUER

CARRERA MIKROTOURBILLONS – REF. CAR5A51.FC6331

The Carrera MikrotourbillonS is not only the world's fastest tourbillon: it is the first ever tourbillon on a 1/100th-second chronograph that can be started and stopped. The MikrotourbillonS has two rotating tourbillon mechanisms visible on its dial face. The first beats at 4Hz (28,800 vph) and controls the watch. The second, the world's fastest tourbillon, controls the 1/100th-second chronograph and runs at 50Hz, meaning it beats at 360,000 vph and rotates at a dizzying five seconds per revolution, or 12 times a minute. Another technical feat: the cage-free mechanism can be started and stopped thanks to its dual-chain architecture, which incorporates one kinematic chain for the watch and a second for the chronograph.

TAG HEUER

MONACO V4 TOURBILLON – REF. WAW2081.FC6348

This piece celebrates the 10-year anniversary of TAG Heuer's revolutionary V4 technology. Entirely crafted and assembled by hand, the Monaco V4 Tourbillon adds an unparalleled dimension to the tourbillon. The masters at TAG Heuer added a micro-belt to drive the tourbillon. This belt ultimately drives the mechanism, resulting in absolute fluidity of its rotation, with no backlash whatsoever. Housed in a space-grade black titanium case, this piece contains four notched micro-thin transmission belts, which remain one of the best-kept secrets in the haute horology world.

VACHERON CONSTANTIN

MALTE TOURBILLON – REF. 30130/000R-9754

Powered by the Vacheron Constantin 2795 caliber, the Malte Tourbillon is stamped with the Poinçon de Genève. This model, which was developed and manufactured entirely in-house, has a power reserve of approximately 45 hours and oscillates at 18,000 vph. The glistening 18K 5N pink-gold case is 38x48.24mm and acts as the perfect frame for the stunning, silvered, sand-blasted dial within. The tourbillon carriage with small seconds is at 6:00 and is accompanied by 18K 5N pink-gold hour markers, a Roman numeral at 12:00 and a black painted minute track.

VACHERON CONSTANTIN

MALTE TOURBILLON OPENWORKED – REF. 30135-000P-9842

The masters at Vacheron Constantin crafted this tonneau-shaped stunner entirely in house. This piece is entirely openworked and the Calibre 2790 SQ beats at the heart with a tourbillon at 6:00. The regulating organ is also complete with small seconds as the central hours and minutes are indicated by hand. Housed in a 950-Platinum case, The Malte Tourbillon Openwork took over 500 hours to construct perfectly. From any given angle, the owner can see a subtle interplay of light and shadows on this beautifully designed piece.

VACHERON CONSTANTIN

PATRIMONY TRADITIONNELLE TOURBILLON 14 DAYS – REF. 8900-000R-9655

This piece houses an exceptional tourbillon and exemplifies the refinement of an understated, round, 18K rose-gold case housing a complication boasting unprecedented performance. The Vacheron Constantin Calibre 2260 beats at the heart of the Patrimony Traditionnelle Tourbillon 14 Days. In order to provide a full 14-day power reserve, this Vacheron Constantin caliber is equipped with four barrels mounted in coupled pairs. The silvered opaline dial provides a view of the tourbillon, while the sapphire crystal caseback exposes the entire movement.

ZENITH

PILOT TYPE 20 TOURBILLON – REF. 87.2430.4035/21.C721

This titanium timepiece with 18K rose-gold bezel plays with asymmetry to achieve a sophisticated architectural choreography reminiscent of vintage aviation instruments. Powered by the self-winding El Primero 4035 mechanical movement, with a power reserve of 50 hours, an intricately constructed tourbillon with integrated small seconds pierces the matte black dial in its upper left quadrant. The date is positioned around the carriage of the gravity-regulating gyroscopic spectacle, cleverly displayed through a revolving aperture. The legible Arabic numerals are made entirely of SuperLumiNova. Oscillating at 36,000 vph, the 381-component caliber also drives a chronograph with 30-minute and 12-hour counters.

Chronographs

Breguet
Type XXII

Pierre Jacques

CEO of De Bethune

"The DB 29 Maxichrono Tourbillon is as complex as it is marvelously simple."

IN 2014, DE BETHUNE WON TWO PRESTIGIOUS AWARDS FOR ITS DB 29 MAXICHRONO TOURBILLON: the "Chronograph Watch Prize" at the Grand Prix de l'Horlogerie de Genève (GPHG) and "Mejor Cronografo" at the Salon International Alta Relojería (SIAR). These new prizes—the titanium DB 28 had already won the "Aiguille d'or" at the GPHG three years previously—recognize the brand's blend of technical innovations and continuous aesthetic development. It is a wager that has paid off for this independent company, founded 13 years ago by David Zanetta, aesthete and collector, and Denis Flageollet, brilliant watchmaker/designer, and helmed today by Pierre Jacques, CEO.

How would you describe the double award-winning DB 29 Maxichrono Tourbillon?

This chronograph is simply superb! Simple, legible, beautiful, refined. It doesn't have several small counters for the seconds and minutes like a traditional chronograph, but rather five hands on the same central axis. That makes for an ultra-legible dial. In addition, a single pushbutton handles all the controls.

Of course, the simpler a watch appears, the more complex its mechanics actually are. We worked on the research and development behind this timepiece for seven years! It is the first chronograph of its kind. De Bethune has secured several patents for the mechanism. We developed an "absolute clutch" mechanism that gives rise to three semi-independent systems controlled by three interdependent column wheels. These facilitate the functioning of the different chronograph counters. It even measures longer intervals, up to 23 hours, 29 minutes, 59 seconds and nine tenths of a second. We have also filed a patent for the flat terminal curve of the balance spring,

DB 29 Maxichrono Tourbillon

Pierre Jacques, CEO of De Bethune

and have developed with this watch a true wrist-worn tourbillon in silicon and titanium. Thanks to these technological advances, we will be able to produce other calibers that are even more precise and high-performance.

The watch is still ultra-legible and very comfortable. The rose-gold case has a soft, rounded silhouette, emphasized by the characteristic cone-shaped De Bethune lugs. The double "officer-type" caseback opens using a monopusher button inserted into the crown at 3 o'clock. One can thus observe the chronograph movement and its splendid finishings, including the mirror-polished steel bridges. The DB 29 Maxichrono Tourbillon is exceptional; we will produce between just 10 and 20 pieces per year.

How do you delineate your collections?

We have two foundational collections. The DB 25 is more classical, with a round case and hollowed-out lugs. The DB 28 has more contemporary characteristics, with articulated lugs and titanium cases, but the same savoir-faire, notably for the finishings. We are also evolving and developing these collections. The DB 27 is the more accessible version of the DB 28. It has the same tilted lugs and titanium case, but with a different caliber, an automatic one.

De Bethune has its own style that pervades all of our models, without adhering to a fixed design code. You will often see cone-shaped lugs or moons in each line, or certain shapes of cases. For example, the DB 29 Maxichrono Tourbillon, which is very classical, uses a case from the DB 15. However, when we want to create something new, for example with the Dream Watch, we give ourselves complete freedom of action.

Do you have a watch that is emblematic of the brand?

Yes, we have several! First of all, the DB 25 QP, with its magnificent curved, guilloché dial, characteristic of De Bethune, and its moonphase. The DB 28 is also emblematic of our work and of our company. It is also the watch I wear; it is light, comfortable and sporty.

What was most memorable for you about the year 2014?

2014 was a year of continuity for us. We presented many watches, notably the DB 29 Maxichrono Tourbillon, the DB 28 Maxichrono, the DB 28 Digitale, the Dream Watch, as well as a new perpetual calendar with the DB 25 QP. They all represent major technical as well as aesthetic innovations, such as the enameled dials. This result, the fruit of years of research, has garnered us a lot of recognition at the GPHG and SIAR for the DB 29 Maxichrono Tourbillon, as well as in Hong Kong with the Spiral magazine award for the Dream Watch 5.

DB 25 QP

De Bethune's Auberson facilities

What are De Bethune's goals for 2015?

We are keeping production low, around 400 watches, and we will continue to propose several new watches, including a DB 28, a first ever De Bethune World Time model, a chronograph and an extremely sophisticated grand complication model.

In terms of distribution, De Bethune is a niche brand; we know our clients and we know our markets. We sell part of our production ourselves. Our watches are also available in about 40 points of sale, mainly in Singapore, the United States and other places in Asia.

What does independence mean for De Bethune?

It does complicate some things, such as access to certain resources, financing or even visibility. These constraints are amply compensated for by total creative freedom. We produce our watches in a very fulfilling work environment.

What should an independent brand look forward to in the next 10 years?

Will there be more or fewer independent brands? That will depend on economic fluctuations, but also on the ability of horology to remain an attractive option for the younger generations. Our grandparents collected pocket watches. Nobody uses those anymore. Even though most of my friends' children wear wristwatches at age 18, that doesn't mean that future generations will. The mechanical watch is not very useful, compared to the quartz watch, which counts extreme accuracy among its advantages. However, it is analogous to the world of cars. One may drive a Toyota Prius and continue to dream of a beautiful Lamborghini. Human beings are not perfectly rational creatures!

Are there more watchmaking connoisseurs these days?

There were already many connoisseurs at the beginning of the 2000s. They are still around, and new collectors are popping up, notably in emerging markets. The information is very readily available, and many clients are looking to understand and really know the watches they are buying.

DB 28 Digitale

Dream Watch 5

Ricardo Guadalupe
CEO of Hublot

"The Big Bang revolutionized Hublot."

IN 2015, HUBLOT CELEBRATES 10 YEARS OF ITS HIGHLY SUCCESSFUL FLAGSHIP COLLECTION, THE BIG BANG. Over the last decade, the brand has established its horological legitimacy, and it now equips the Big Bang UNICO (evolution of the Big Bang) with in-house movements, as well as the ultra-thin Classic Fusion. This year, Hublot will also enlarge its production capacities and enhance the Big Bang collection with a new house-made complication.

What does the Big Bang represent for Hublot?
The Big Bang made Hublot what it is today. This model singlehandedly revolutionized the brand. It is an "accelerated" adaptation of a famous classic Hublot watch from 1980—20 years of evolution condensed into one model. The Big Bang represents the essence of Hublot, which is the Art of Fusion in watchmaking, a blend of tradition and a vision of horology's future. It combines materials such as rubber, which we were the first to introduce in horology, with ceramic, Kevlar and many other materials.

The public identifies Hublot with the Big Bang. Is this an advantage or a disadvantage?
The Big Bang has brought us much success, and we have a lot to thank it for! It created a shock wave and allowed us to relaunch Hublot. The most prestigious watch brands are all identified with one iconic watch, an absolute reference that ensures their longevity. So Hublot has the Big Bang.

How do you respond to the allegations that the Big Bang resembles another well-known watch?
These accusations should have been made back in 1980, when Hublot released its first watch. With the Big Bang, we simply respected the design codes of that original watch. Like that other famous watch, it resembled a boat's porthole, so there were, of course, similarities between the two interpretations. Today, however, the Big Bang has earned its own reputation, and the comparison has faded.

Big Bang 2006

Ricardo Guadalupe
CEO of Hublot

Big Bang Unico All Black

Classic Fusion Power Reserve

Big Bang Ferrari

In your opinion, which Big Bang model best represents the collection?

The All Black was created in 2006. Since then, we have presented a limited edition each year. Formerly, horology was focused on gray and gold. With the All Black model, we turned black into a high-end standard. We pushed the concept to extremes, with much success.

What is your favorite Big Bang?

The very first one, in steel and ceramic. This original version is also our best seller, even though we created it in 2004 and launched it in 2005, ten years ago! Eventually we will replace it with the Big Bang UNICO. It is the same model, but with our own movement.

How will you develop the Big Bang collection in the future?

Hublot will maintain our production output of about 40,000 watches per year, which is quite exclusive. To move toward the higher end, we will increase the horological value of our timepieces. That is why we will round out the collection with Big Bang models worth $50,000-$100,000 USD. Though we already offer tourbillons and minute repeaters, there are still several complications we don't do, such as perpetual or annual calendars, GMT indications or alarms. This year, we are presenting a perpetual calendar chronograph, with a UNICO base movement.

Could the Big Bang eventually run out of steam?

The key to success will be to continue its evolution, with new forms and functions. Just as Porsche has done with its 911 model! We have launched a second line, the Classic Fusion. We are also performing experiments in design and exploring new forms, such as the tonneau case. All these variations conserve Hublot's DNA. But the typical Hublot watch, such as the Big Bang or the Classic Fusion, will stay round.

How does the Classic Fusion relate to the Big Bang?

If the Big Bang was a shock wave, the Classic Fusion more faithfully reflects the Hublot of 1980. It has emerged within a trend toward classic, chic and elegant looks. It is more suited to the needs of a more traditional clientele. However, this Hublot interpretation of a classic watch maintains a strong brand identity. It has a rubber strap and uses ultra-thin movements with a high level of watchmaking sophistication. We believe that high-end watchmaking will survive through the value that it represents. That is why we emphasize our in-house movements and sometimes reveal them by choosing to omit the dial.

Hublot offers just two main collections and many limited series. Why is that?

The heart of our line, with a hundred different models, remains the same: watches in steel and ceramic, gold and ceramic or ceramic alone, released in several different sizes. At the same time, each year we offer new variations in limited editions, which suits the tastes of some of our clientele, who already own our basic models. This division also shows up in our points of sale, which offer a base of 30 to 60 watches and an assortment of limited series.

These limited series are notably a part of your partnership with Ferrari. How successful has this alliance been?

It has been a great success for both Ferrari and Hublot! We are partnered with the enormous prestige of Ferrari, a worldwide luxury reference. Before us, other watchmakers had not been able to make this collaboration work. We built the partnership from the ground up, based on a free exchange between the teams. Our objective, which we achieved, was to sell very high-end Hublot-Ferrari special editions to the owners of Ferrari automobiles! We are present at all the Ferrari events in the world—about 200 last year—from the smallest to the largest, such as the 60th anniversary of Ferrari in the USA, celebrated in Beverly Hills with a gathering of over 1,000 Ferraris. And every Ferrari representative has a Ferrari clock in the garage or showroom. Hublot is also present on the F1 car driven by Team Ferrari, which brings us some nice visibility. The five-year contract will be renewed.

Finally, Hublot has a new brand ambassador, Bar Refaeli. Which clientele are you targeting?

Bar Refaeli allows us to approach women in a brand-new way, with a photo shoot done by a famous Chinese photographer. Our goal is to sell 30% of our watches to women, as opposed to 25% today. Our creations, which are highly innovative, brightly colored and creative, appeal to women who already own classic watches and want some variety.

Aldo Magada
CEO of Zenith

Aldo Magada

CEO of Zenith

"The El Primero Chronomaster 1969 sums up what Zenith stands for, both aesthetically and technically."

ZENITH OPENED A NEW CHAPTER IN 2014, WITH THE ARRIVAL OF ALDO MAGADA AS CEO. In 2015, the watch manufacturer celebrates its 150th anniversary. Its new models continue the approach of refocusing the collections on three lines. The flagship collection is the legendary El Primero, equipped with its famous eponymous movement, which is accurate to one-tenth of a second. The El Primero Chronomaster 1969 model stands out as THE emblematic face of Zenith. The sporty Stratos and the classic Elite collections round out the brand's range. Zenith also continues to offer highly exclusive pieces, and even the option of creating personalized watches.

El Primero Chronomaster 1969

How would you define Zenith?

Above all, Zenith is defined by the high-frequency El Primero chronograph. It is the only series-produced chronograph of its kind on the market. Within the El Primero collection, Open represents the most recognizable, most important iteration of the concept. It's a very classic watch, but with a twist: the dial-side aperture that reveals the escapement.

Zenith has two other pillars: Stratos in the sport line and Elite for more refined models focused on hours, minutes and seconds. Elite replaces the former Captain collection. All the Captain chronographs will be integrated into the El Primero collection. So the Elite collection concentrates on three-handed models, with of course supplemental complications, such as the date, power reserve indication or moonphase. This kind of model is tremendously popular, particularly among women.

Which watch best represents Zenith?

The El Primero Chronomaster 1969! We would like it to be the face of Zenith on the market, as it clearly sums up what Zenith stands for, both aesthetically and technically. This iconic model brings together the historic colors of the first El Primero chronograph and Zenith's trademark dial-side aperture revealing the movement. The classic 42mm case suits men and women. The bezel is rather slim, providing a more comprehensive view of the dial, which we always finish with beautiful decorations. We are also improving the angles of the lugs, to make them even more comfortable for smaller wrists. We regard finishes and details as extremely important, even if they are not immediately visible.

How is the technical aspect of the El Primero Chronomaster 1969 representative of Zenith?

In 1969, Zenith presented the first integrated automatic chronograph movement. It is still the world's most precise series-produced mechanical movement, as its 36,000 vph frequency allows it to measure times to one tenth of a second! It has since become legendary, and powers all the El Primero Zenith chronographs. This proud lineage includes the El Primero Chronomaster 1969, released in 2012. It houses the El Primero 4061 caliber, a recent El Primero chronograph movement whose regulating organ has been moved to 10 o'clock so as to be visible through the dial aperture. This automatic caliber has been entirely designed, perfected and produced in-house. Its technical advantages include an over 50-hour power reserve and its lever and escape-wheel are made of silicon to ensure enhanced performance, especially in terms of longevity. Finally, it is water resistant to 100m.

Academy Georges Favre-Jacot

What are you presenting at Baselworld this year?

Among various new models, we have a new, sporty Stratos watch, in the same spirit as its predecessors. In the Elite collection, we have a slightly larger men's model, with a diameter of 42mm. It is equipped with a new, automatic, three-hand Elite movement and it will be the base for future developments in the collection. In January, we also presented in Geneva a second Zenith Academy Christophe Colomb Hurricane Grand Voyage—an exceptional model.

The famous gyroscopic escapement developed by Zenith

So you will continue to produce higher-end watches?

Absolutely! Even though our typical price point is between 9,000 and 12,000 Swiss francs, we will always create pieces of haute horology, particularly with the decorated El Primero cases.

When I see all the projects we have in the pipeline at the Manufacture, I know that we can serve a certain clientele in search of exclusive watches. We are building up our haute horology facilities with a dedicated workshop. Our clients can order bespoke, completely personalized watches. Someone who orders a watch in this way can communicate with and collaborate with a single watchmaker who will construct the entire watch, as well as providing service after the purchase.

What about research and development?

In terms of research in haute horology, the gyroscopic escapement on Zenith's Grand Voyage constitutes a real answer to a known problem. It is an intelligent and elegant solution that prevents gravity from exerting an undue influence on the functioning of the regulating organ. At the end of 2014, we opened the celebrations of our 150th anniversary with the Academy Georges Favre-Jacot, featuring a large aperture across the upper part of the dial providing a chance to observe the fusee-and-chain transmission with the barrel at 10:30, as well as the fusee around which the chain wraps itself and which appears at 1:30.

In terms of displays, we are headed toward more new features and we wish to create interesting and different experiences for the consumer.

Zenith Hurricane chain movement

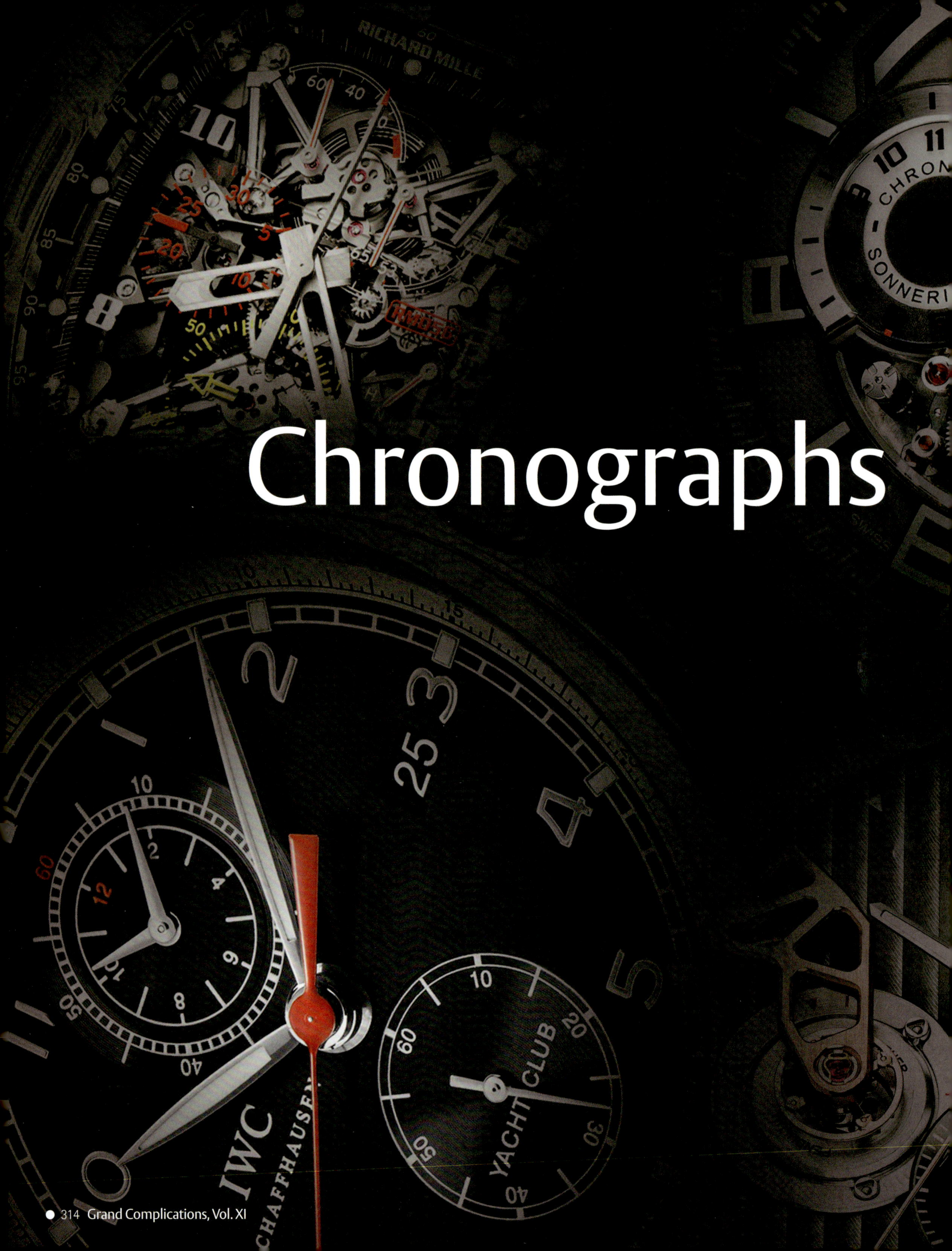

Chronographs

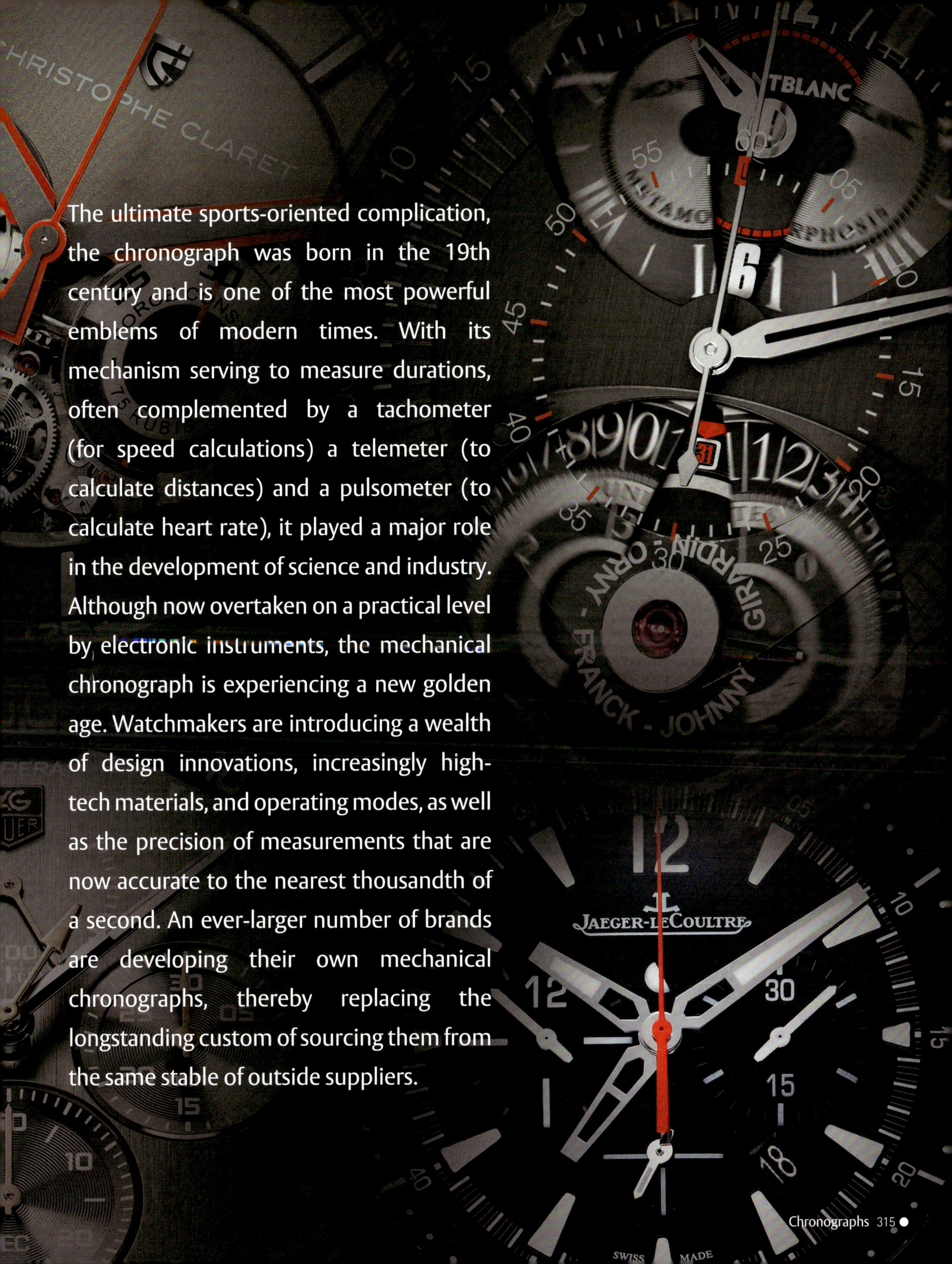

The ultimate sports-oriented complication, the chronograph was born in the 19th century and is one of the most powerful emblems of modern times. With its mechanism serving to measure durations, often complemented by a tachometer (for speed calculations) a telemeter (to calculate distances) and a pulsometer (to calculate heart rate), it played a major role in the development of science and industry. Although now overtaken on a practical level by electronic instruments, the mechanical chronograph is experiencing a new golden age. Watchmakers are introducing a wealth of design innovations, increasingly high-tech materials, and operating modes, as well as the precision of measurements that are now accurate to the nearest thousandth of a second. An ever-larger number of brands are developing their own mechanical chronographs, thereby replacing the longstanding custom of sourcing them from the same stable of outside suppliers.

THE PATERNITY OF THE CHRONOGRAPH WAS LONG ASCRIBED TO FRENCH WATCHMAKER NICOLAS RIEUSSEC, who in 1821 created an instrument that placed a drop of ink on a dial at the beginning and end of each measurement.

However, the year 2013 brought a complete upheaval of this established order with the discovery of an 1816 timepiece signed by Louis Moinet—a compteur de tierces or a counter displaying sixtieths of a second by means of a central hand, seconds and minutes on two separate subdials, as well as hours on a 24-hour dial—complete with start, stop and reset functions controlled via two pushers.

The year 1838 saw the invention of the split-seconds chronograph (see "How does it work?"), and the first system to reset the hands to zero gained widespread popularity around mid-century. Now equipped with its three main functions (start, stop, reset), the chronograph would become very useful in several domains (science, technology, transportation, the military), and it is not by chance that its rise coincided with that of the industrial society and competitive athletics. During this period the chronograph would also gain several new scales for various measurements: tachometer (speed), telemeter (distance), pulsometer (pulse), etc.

The first wristwatch chronographs made their appearance in 1915. They were equipped, as were the pocket chronographs, with a single pushbutton for all three functions, usually housed in the crown. Breitling designed the now dominant chronograph configuration by adding the first independent pushbutton, and then in 1934, another one for resetting the chronograph. In 1937, the Dubois Dépraz manufacture presented a movement whose chronograph command operated using a cam system—less expensive and less difficult to produce than the traditional column wheel. The system would slowly be adopted by most brands.

In July of 1969, the Speedmaster, from Omega, would accompany American astronauts on their first visit to the moon. That year was also marked by the appearance of the first automatic chronograph movements. The movement developed by the Büren/ Breitling/Heuer-Leonidas threesome in collaboration with Dubois-Dépraz was a cam movement featuring a modular construction, with an off-centered microrotor (oscillating weight); the one from Zenith, the famous El Primero, was an integrated movement with a column wheel in its central rotor. Thanks to its frequency of 36,000 vph, it was also the first chronograph movement capable of measuring times to one-tenth of a second.

The advent of quartz movements, which were much more precise, was nearly the deathblow for the mechanical chronograph, but the renaissance of mechanical horology in the late 1980s signaled its triumphant return. Let us remember that it is very rare for watch brands to design and produce their own chronograph movements. A large percentage of the chronographs on the market today house an ETA 7750 Valjoux movement, reworked to a greater or lesser degree and sold with differences in price that can be hard to justify... But the situation is beginning to change. The public, better informed on the subject than previously, is more than willing to pay a premium for a high-end chronograph, but on the condition that it is an exclusive movement—or at least somewhat original. The Swatch Group's announcement that it would stop selling ETA movements has also inspired competing brands to look for alternative solutions. For several years, we have been seeing an increase in the number of "in-house" chronographs featuring interesting innovations.

HOW DOES IT WORK?

The chronograph functions using a complex system of levers, springs and wheels. The chronograph mechanism is "grafted" onto the movement's base. When the "start" pushbutton is pressed, the chronograph wheel joins the wheel for the seconds, and the hand starts ticking. When the chronograph is stopped, the chronograph wheel separates from that of the seconds, and the chronograph hand stops again. The reset function is effectuated by a system of a hammer on a spring and a "heart."

SPLIT-SECONDS CHRONOGRAPHS

Split-seconds chronographs are equipped with an additional seconds hand, which can be stopped independently to measure a "split" time or a reference time. They are recognizable by the additional pushbutton they bear. Activating the pushbutton is all that is needed for the split-seconds hand to catch up with the main chronograph hand, so that the two may continue running together.

COLUMN-WHEEL AND CAM MOVEMENTS

There are two kinds of chronographs with respect to the "control group" governing the start, stop and reset functions. The chronographs considered the most prestigious are equipped with a column-wheel, a complex organ that uses a system of slots and tabs to govern the movements of the levers. Most current mechanical chronographs are equipped with a cam system of simpler construction.

MODULAR AND INTEGRATED CONSTRUCTIONS

In modular constructions, the chronograph mechanism is totally independent of the base movement. It is mounted on an additional plaque affixed to the dial side of the movement. This solution allows the watchmaker to link a chronograph mechanism to an existing movement, without having to develop a new caliber. In integrated constructions, the chronograph components are placed directly on the movement, on the bridge side. Column-wheel chronograph movements are the result of integrated constructions.

FREQUENCY AND PRECISION

The precision of time measurement and display of a mechanical chronograph depend on the frequency of its regulating organ, i.e. the number of vibrations per hour (one oscillation equals two vibrations) of the balance-spring. The higher the frequency, the smaller units of time that can be measured.

Vibrations per hour	Frequency (in Hz)	Smallest measurable unit of time
18,000	2.5 Hz	1/5 of a second
21,600	3 Hz	1/6 of a second
28,800	4 Hz	1/8 of a second
36,000	5 Hz	1/10 of a second
360,000	50 Hz	1/100 of a second

FLYBACK FUNCTION

The flyback function allows the wearer to instantly start a new round of timing while the chronograph is already operating. For this, one need only activate the reset pushbutton, without stopping the hands first; all three chronograph hands will immediately return to zero to start another measurement.

JUMPING SECONDS

The term "jumping seconds" refers to a setup in which the second hand leaps around the dial in one second—usually in four or five smaller steps. This system is a way to measure and display times of one-fourth, one-fifth—even one-eighth or one-tenth of a second.

CHRONOGRAPH AND CHRONOMETER

The word "chronometer" is often used incorrectly to designate a chronograph, but they are two very different things. A chronometer is a watch whose movement has received an official certificate attesting to its extreme precision. This certificate is awarded by the COSC (Swiss Official Chronometer Testing Institute) after two weeks of very rigorous testing. A chronograph may be a certified chronometer, but it is by no means certain that any given chronometer will also be a chronograph.

A Reinvented Classicism

Chronographs have long been distinguished by their rather classic aesthetic, with three small symmetrically placed subdials. This traditional look has now been swept aside by many models that reinvent the ways in which we measure time. The chronograph universe has also been marked by the appearance of new materials that allow the watches to become lighter, more robust or more technically sophisticated, as well as shaking up aesthetic codes.

Jaeger-LeCoultre, *Master Compressor Chronograph Ceramic*

The ceramic case of the Master Compressor Chronograph Ceramic by Jaeger-LeCoultre houses a Manufacture Jaeger-LeCoultre automatic-winding movement. The patented compression key system enhances the reliability of the crown seal.

Graham
Chronofighter Oversize Superlight Carbon

With its 47mm black carbon nanotube composite case, Graham's Chronofighter Oversize Superlight Carbon chronograph weighs in at less than 100g.

Breitling
Chronomat Airborne 44
A special series of Breitling's flagship Chronomat collection, created in 1984 for the elite Frecce Tricolori flight team, the Chronomat Airborne 44 revives the look of the original movement. It is equipped with Manufacture Breitling Caliber 01.

Blancpain
Chronographe Villeret Pulsomètre
In addition to its flyback function, which allows the wearer to start a new timing operation without needing to stop and reset the chronograph to zero, the Villeret Flyback Pulsometer Chronograph displays a pulsometric scale enabling fast and easy heart-rate measurements.

Maurice Lacroix
Masterpiece Le Chronographe Squelette
As its name implies, the automatic-winding Masterpiece Le Chronographe Squelette by Maurice Lacroix has an entirely skeletonized chronograph, providing admirable views of its multi-layered structure.

Zenith, *El Primero Lightweight*
A high-tech version of its El Primero Striking 10th chronograph (with a central chronograph seconds hand that displays tenths of a second), the Zenith El Primero Lightweight combines a titanium and silicon movement with a carbon fiber case.

Rolex
Oyster Perpetual Cosmograph Daytona
In 2013, Rolex celebrated the 50th anniversary of the Oyster Perpetual Cosmograph Daytona—one of the most legendary chronographs of all time—by launching a new version with a platinum case, chestnut brown Cerachrom bezel and glacier blue dial.

Omega
Speedmaster Dark Side of the Moon
The only watch to have been worn on the Moon, hence its "Moonwatch" nickname, the famous Omega Speedmaster notably comes in a "Dark Side of the Moon" version featuring a black ceramic case and dial along with an Omega Co-Axial escapement.

Cartier, *Tank MC Chronographe*
The famous Tank de Cartier has been interpreted in an "MC Chronograph" version, with a Cartier movement featuring a column wheel, vertical coupling clutch and twin-barrel automatic winding guaranteeing perfectly constant long-term mainspring torque.

Omega, *Seamaster Bullhead*
Inspired by a 1969 model, the Seamaster Bullhead by Omega opts for originality with an asymmetrical shield-shaped case and a crown at 12 o'clock framed by two chronograph pushers. The 6 o'clock crown rotates the inner bezel.

Regatta Chronographs

Models described as "regatta chronographs" possess a notable countdown system that marks off the last five or ten minutes before the race begins, before starting the timing operation as such. With the growing popularity of sailing competitions such as the America's Cup, these models are getting new wind in their sails.

Louis Vuitton
Tambour Spin Time Régate
On the Tambour Spin Time Régate by Louis Vuitton, the five minutes of the countdown used in the America's Cup are indicated by five rotating cubes switching from red to black.

Rolex
Oyster Perpetual Yacht-Master II
Rolex has reaffirmed its place in the sailing world by launching the Oyster Perpetual Yacht-Master II, which boasts an ingenious bezel/crown/pushbutton system that follows the countdown using a large triangular hand.

Alpina, *Yacht Timer Regatta Countdown Automatic*
An active brand in the field of nautical sponsoring, Alpina has equipped its Yacht Timer Regatta Countdown Automatic with a five-to-ten-minute countdown complete with fluorescent orange "start" indicator.

Corum, *Admiral's Cup AC-One 45 Regatta*
Recognizable by its twelve-sided bezel, the Admiral's Cup AC-One 45 Regatta from Corum has a countdown function that can be adjusted up to ten minutes via the crown, with a digital aperture-type display ensuring optimal readability.

Panerai
Luminor 1950 Regatta 3 Days Chrono Flyback Automatic Titanio
In addition to its pusher-adjustable countdown system, the Luminor 1950 Regatta 3 Days Chrono Flyback Automatic Titanio by Panerai displays a tachometric scale in nautical knots serving to measure boats' speed.

AUDEMARS PIGUET

ROYAL OAK OFFSHORE CHRONOGRAPH – REF. 26401RO.OO.A002CA.01

The Royal Oak Offshore Chronograph's automatic-winding Manufacture caliber 3126/3840 is comprised of 365 parts, with 59 jewels, and oscillates at 21,600 vph. The 22K gold oscillating weight contains ceramic ball bearings. All of these internal components fuel the chronograph that the wearer sees on the black dial with the signature Audemars Piguet "Mega Tapisserie" pattern. The 18K pink-gold case and black ceramic bezel are accompanied by a screw-locked crown and glareproof sapphire crystal and caseback. This ultra-durable model is water resistant to 10atm.

AUDEMARS PIGUET

ROYAL OAK OFFSHORE CHRONOGRAPH – REF. 26402CB.OO.A010CA.01

This sleek, brightly colored version of the classic Audemars Piguet Royal Oak Offshore Chronograph is complete with a 44mm white ceramic case and bezel. The titanium pushpiece guards and links give the snowy-toned wristwatch a sense of strength and bring out the light, silvery hue of the "Mega Tapisserie" dial. The 12-hour counter at 6:00, 30-minute counter at 9:00 and small seconds counter at 12:00 are all navy blue, as is the tachometer. The central seconds hand has a red tip to match the hands on the hour and minute counters.

BELL & ROSS

BR-X1 SKELETON CHRONOGRAPH

The skeleton chronograph movement of the BR-X1 combines haute horology finishes and extreme lightness. The "X" in its name refers to the shape of its upper bridge, as well as the code name of NASA's experimental projects. The 45mm titanium case, which sports rubber inserts and toggle pushbuttons, reveals the movement's balance through a tinted sapphire crystal opening in the caseback—a Bell & Ross signature. Crafted in gray-tinted mineral glass, the transparent dial allows a generous view of the exceptional movement. The watch displays hours, minutes and small seconds, a skeletonized date at 6:00 and a chronograph with 60-second and 30-minute accumulators. The 30-minute timer on the chronograph function is positioned at 9:00 and features an ultra-light aluminum disc resembling the blades of a turbine.

BREGUET

TYPE XXII – REF. 3880-BR-Z2-9XV

Crafted in 18-karat rose gold, the 44mm case of the Type XXII bears rounded horns, a bidirectional rotating bezel and a finely fluted caseband. Its screw-down crown contributes to a water resistance of 100m. The warm brown dial, signed Breguet, boasts luminous hands and hour markers to complement the chapter ring with Arabic numerals. A red central chronograph seconds hand operates on a 30-second basis, as does a small seconds subdial at 9:00. At 6:00, the dial sports a second time zone indicator and date indication. A 24-hour indicator adds to the clarity of the display at 3:00. The chronograph movement, Cal. 589F, includes a flyback function and 40 hours of power reserve. The silicon escapement operates at a frequency of 10 Hz.

CARL F. BUCHERER

PATRAVI CHRONODATE – REF. 00.10624.08.33.21

This chronograph is housed in a stainless steel case that measures 42mm in diameter and 14.1mm in thickness. The black dial has eight polished hour markers, which dance around the counters and indications in the center. The large date at 12:00 is balanced by the 12-hour counter at 6:00, 30-minute counter at 9:00 and small seconds counter at 3:00. There is also a central chronograph seconds hand. The Patravi Chronodate is powered by the self-wound CFB 1956 caliber and has a power reserve of 42 hours.

CHRISTOPHE CLARET

DUALTOW – REF. MTR.CC20A.012

The DualTow features a planetary-gear, monopusher chronograph with an original striking mechanism signaling start, stop and reset function. It also contains a tourbillon escapement and two spectacular rolling belt indicators running parallel on each side of the dial to indicate hours and minutes. The movement is composed of 568 components and 65 jewels. DualTow is sold as a limited edition of 68 unique pieces.

CHRISTOPHE CLARET

KANTHAROS – REF. MTR.MBA13.905

This mono-pusher chronograph with striking mechanism and constant force escapement asserts its own identity and reinvents the measurement of timing events. The automatic-winding chronograph's cathedral gong, visible at 10:00 audibly chimes with each change of function. In addition to this innovative complication, Kantharos is also equipped with a constant-force escapement at 6:00. This mechanism is a key component that considerably reduces timing variations. The energy delivered to the escapement remains constant from beginning to end of the power reserve. Precision is the overriding goal. The perpetually moving mechanism exercises a hypnotic visual effect, which may be admired beneath a meticulously chamfered sapphire bridge revealing the full extent of the master-watchmaker's know-how.

DE BETHUNE

DB28 MAXICHRONO – REF. DB28MCRZN

De Bethune's DB28 Maxichrono combines the old and the new in both its design and technical construction. The rose-gold case contrasts sharply with the blackness of its floating lugs in hand-polished, black zirconium. A 21st-century take on time readoff uses contemporary skeleton hands and black numerals in a modern typeface. Elapsed times are recorded more conventionally, with traditional blued steel hands and numerals borrowed from marine chronometers. Inside, the DB28 Maxichrono remains a modern chronograph, with a button paired with the crown controlling the five hands mounted coaxially in the center of the dial. On the back, the window on the calibre DB2030 displays the architectural layout of the movement's bridges in rose gold and polished steel. This model is available in a limited edition of 20 pieces.

F.P. JOURNE

CENTIGRAPHE SPORT IN TITANIUM

F.P. Journe introduces the Centigraphe Sport with a case in grade-5 titanium, a high-tech and ultra-light material of anthracite-gray colour with satin finish. In a quest for performance and lightness, the hand-wound haute horology movement and dials are made of aluminum alloy and this watch weighs between 60 and 70g depending on the model. The Centigraphe Sport indicates elapsed times from a 100th of a second to 10 minutes, visible on 3 dials, each with a time scale in red and a tachometer scale in black. The ergonomic chronograph mechanism is patented.

FRANCK MULLER

CHRONOGRAPHE AUTOMATIQUE MARINER

The Chronographe Automatique Mariner's radiant rose-gold case comes complete with a black crown and pushbuttons. These dark external parts not only match the natural rubber strap but the black sunray guilloché dial as well. The dial is spruced up by royal blue appliqués and white Arabic numerals and hands. The color scheme and compass décor of the dial give the model a vintage navigational feel. The chronograph is to 1/5 of a second with a 30-minute counter at 3:00 and a central seconds hand. The small seconds at 9:00 and date at 6:00 also guarantee exactitude for the Mariner's owner.

FRANCK MULLER

CONQUISTADOR GRANDPRIX

Bright, bold and dashing, the Conquistador Grandprix Chronograph shows Franck Muller's adventurous side. With an automatic-winding FM 700 caliber, a power reserve of 48 hours and a platinum 950 oscillating mass, this model is as powerful on the inside as it is on the outside. The black dial has a flame pattern with Arabic numerals painted in red relief. The titanium and ergal curved case of this Conquistador model is complemented by a black alligator strap with red stitching. Black and red make for an exceptionally strong-looking timepiece that is extremely functional as well. This can be attested for by its ultra-reliable bi-wheel chronograph.

GLASHÜTTE ORIGINAL

SENATOR CHRONOGRAPH PANORAMA DATE – REF. 37-01-01-05-30

The new Caliber 37-01 column wheel chronograph is introduced at the heart of the Senator Chronograph Panorama Date. The model is complete with a red-gold case and lacquered silver grained dial. This piece displays an elegant 12-hour chronograph counter at 12:00, while the Panorama Date display with black numerals on an ivory ground is positioned at the center of the lower half of the dial. The small seconds counter with a power reserve indicator in black and white appears at 9:00 while the 30-minute chronograph counter is at 3:00. The hands are all blued steel and the piece is complete with a power reserve of 70 hours.

GLASHÜTTE ORIGINAL

SEVENTIES CHRONOGRAPH PANORAMA DATE – REF. 37-02-03-02-70

The Seventies Chronograph Panorama Date is a sporty chronograph with distinctive vintage appeal. The new Caliber 37-02 was conceived specifically as an integrated chronograph movement and is featured in this model. The masters at Glashütte Original combined central stop seconds, 30-minute and 12-hour counters with a flyback mechanism, a small seconds counter with a power reserve indicator of 70 hours. There is also the signature Panorama Date Display. The column wheel has been designed to be as simple and compact as possible. Mounted within the Seventies Collection's 40x40mm stainless steel case, the moderate size and simplicity enhance stability.

GUY ELLIA

JUMBO CHRONO

The Jumbo Chrono, the first round men's watch created by Guy Ellia, is based on an interesting technological approach demonstrating that it is possible to combine aesthetic balance with masculine detail and a fine mechanical movement. The Jumbo Chrono boasts an exceptional 50mm size that is rarely seen in the world of luxury watches. The Jumbo Chrono's rhodium-plated rotor is unique and the impressive discovery dial reflects a strong graphic design: sweep-seconds chronograph in the center, hour and minute at 12:00, date at 2:00, 30-minute counter at 3:00, seconds at 6:00, and 12-hour counter at 9:00. This model is shown in black gold, but is also available in white or pink gold.

HUBLOT

BIG BANG DARK JEAN CERAMIC – REF. 301.CI.2770.NR.JEANS14

Thanks to a cutting-edge technical development, designer grade genuine dark blue denim is used as a covering on the dial of the iconic Big Bang Chronograph 44mm, after having undergone a complex process to harden the denim and make it easier to work. The timepiece is powered by an HUB4100 self-winding chronograph movement with 252 components (28,800 vph). Water resistant to 100m for all types of swimming activity, this watch is available in a limited edition of just 250. Made entirely from micro-blasted black ceramic, the model is completed by a strap in denim sewn onto black rubber for comfort, flexibility and strength.

REF. 341.SL.5199.LR.1905.POP14

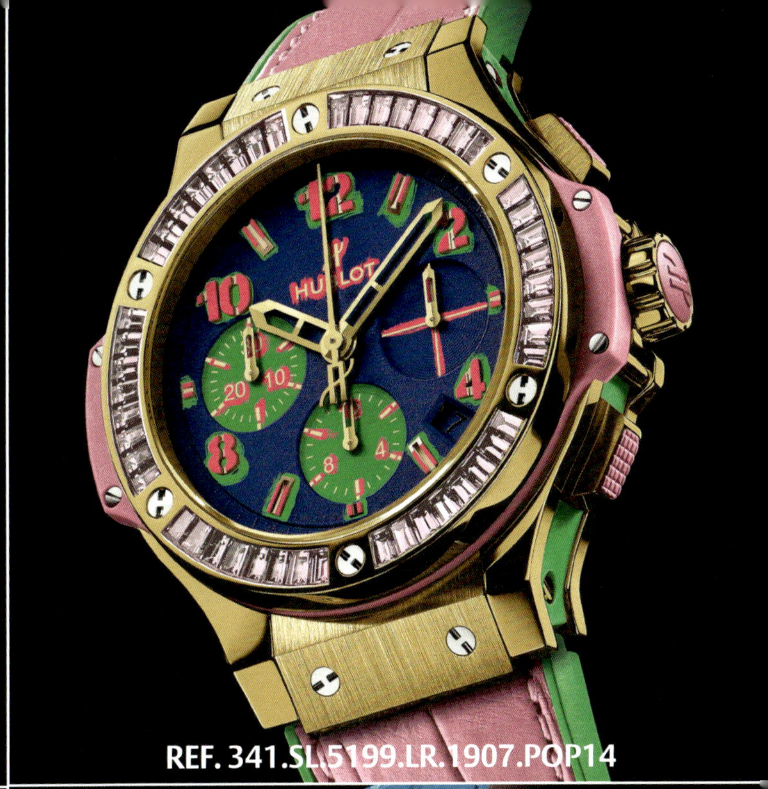

REF. 341.SL.5199.LR.1907.POP14

REF. 341.SL.5199.LR.1922.POP14

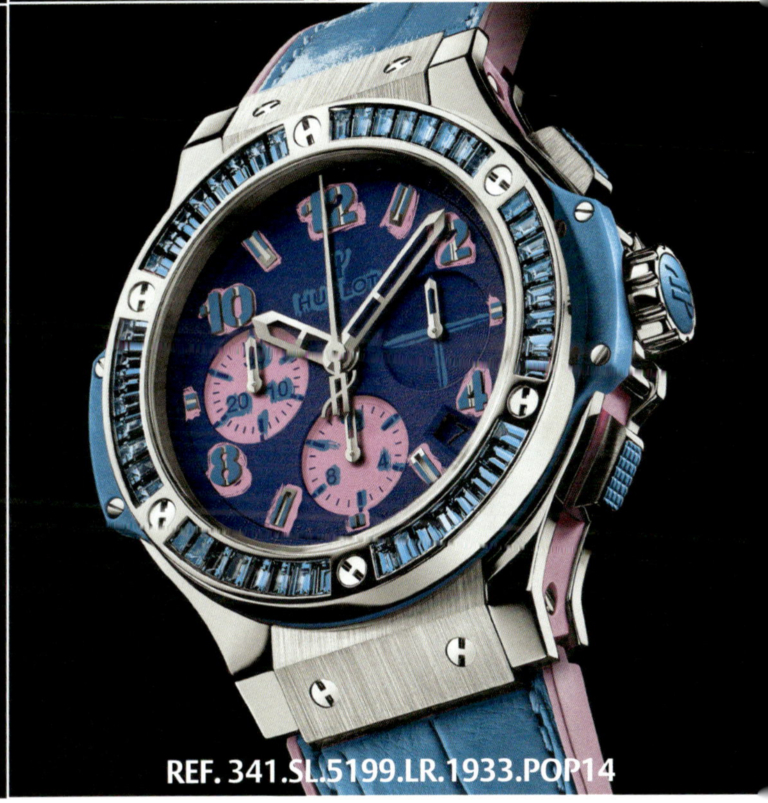

REF. 341.SL.5199.LR.1933.POP14

HUBLOT

BIG BANG POP ART

The playful ethos of Pop Art inspires Hublot's Big Bang Pop Art collection, which boasts four versions that burst with seriously vibrant color: pink, blue, purple, apple green. Two models in gold and two in steel take on the iconic Big Bang 41mm, featuring a bezel set with 48 baguette precious stones: topazes for the blue, amethysts for the purple, sapphires for the pink and tsavorites for the green, and dials with multi-colored counters reminiscent of a painter's palette. The ensemble is completed by a strap in colored alligator sewn onto tone-on-tone rubber. Inside beats the HUB 4300 movement, a self-winding mechanical chronograph with a date window at 4:00. With a power reserve of over 40 hours and water resistance to 100m, the Big Bang Pop Art is durable and reliable as well as shockingly stylish.

HUBLOT

BIG BANG UNICO ALL BLACK – REF. 411.CI.1110.RX

The Big Bang Unico is now available in an All Black version as a limited edition of just 500 numbered pieces, in line with all of Hublot's All Black series. The 45.5mm watch embodies Hublot's unique "invisible visibility" concept. Other signature Hublot elements include the iconic Big Bang design and the in-house movement—the UNICO, a chronograph movement with a column wheel and double clutch, completely designed, developed and produced in the Hublot workshops. The movement is revealed thanks to a skeleton dial featuring redesigned, larger indexes and Arabic numerals that have been hollowed out and filled with black SuperLumiNova for the All Black version. The dial also has two counters: one for the minutes at 3:00 and the other at 9:00 for the small seconds; the date window is positioned at 3:00. The optimized strap attachment features the famous "one click" system that allows the wearer to switch among a wide choice of straps.

HUBLOT

BIG BANG UNICO CARBON – REF. 411. QX.1170.RX

The design of the Big Bang Unico, here shown in carbon fiber, features a powerful 45.5mm diameter, with a bezel secured by 6 H-shaped screws. The optimized strap attachment features the famous "one click" system that allows the wearer to switch among a wide choice of straps. This flyback chronograph is unique in the watchmaking industry, and features two pushbuttons and a date indication. Its development required a completely original approach: the horizontal double clutch with integrated column wheel on the dial side for the central 60-second counter and the small 60-minute counter positioned at 3:00. This simplification means greater reliability and flexibility in the chronograph's START/STOP/RESET functions. The movement's 330 components benefit from high precision and reliability. Its frequency of 28,800 vph and immaculate finish are also a guarantee of high chronometric quality.

HUBLOT

KING POWER SPECIAL ONE – REF. 701.NQ.0137.GR.SPO14

In honor of The Special One—soccer coach José Mourinho—who recently rejoined the prestigious ranks of Hublot ambassadors, Hublot has unveiled the aptly named King Power Special One. With a strapping diameter of 48mm and a sleek, sporty design, the King Power is not your average watch. Blue is Mourinho's signature color, and this watch's attention-getting dark blue carbon fiber is the result of an unprecedented technical process. The color is directly integrated in the fibers arranged in the mold, and the whole assembly is then pressurized. The color is thus within the material. The visual effect is emphasized by the blue molded bezel and several accents in the same color, as well as a navy blue alligator strap. Mourinho's signature also appears on the back of the watch. This exclusive model comes in titanium or King Gold.

LONGINES

THE LONGINES AVIGATION WATCH TYPE A-7 – REF. L2.799.4.53.0

The racy elegance of The Longines Avigation Watch Type A-7, a single pushbutton chronograph, immediately catches the eye. The steel case has an impressive diameter of 49mm and houses an L788 caliber, a column-wheel movement developed exclusively for Longines. The dial is angled at 50° to the right and features large white Arabic numerals, all of which make up the exceptional character of this timepiece. Its sleek appearance is enhanced by the tachometer scale around the outer edge of the dial and by the Breguet minute and hour hands. The whole is mounted on a black alligator strap. The intricate workings of this exceptional piece can be admired through the hinged cover on the sapphire crystal back.

LONGINES

THE LONGINES COLUMN-WHEEL SINGLE PUSH-PIECE CHRONOGRAPH REF. L2.774.8.23.X

The Longines Column-Wheel Single Push-Piece Chronograph models take their inspiration from the first chronograph wristwatches developed by Longines in 1913. With its 40mm diameter, this rose-gold model is fitted with the L788 caliber. This column-wheel single pushpiece chronograph movement has been developed by ETA exclusively for Longines. Its white dial is adorned with painted black numerals and a red "12," in a direct reference to the dials of the period. It is completed by two counters, at 3:00 and 9:00, together with a date aperture at 6:00. Blued hands harmoniously complete the whole.

PATEK PHILIPPE

MEN GRAND COMPLICATIONS – REF. 5950/1A-011

One of the salient features of the new Patek Philippe split-seconds chronograph is its steel case, which is thinner than that of many ordinary three-hand watches. This is due to the exquisite CHR 27-525 PS movement, endowed with the Patek Philippe Hallmark, with which the manufacture set a new global record in 2005 for the thinnest column-wheel chronograph caliber ever made. The rose-gold dial of this model sets off the deep gloss of the polished steel case. The crisp railway track minute scale emphasizes the masculine, technical look of the split-seconds chronograph. The four dial corners between the circular minute scale and the cushion-shaped bezel are decorated with fine black-lacquered engravings that add a touch of graceful elegance. Short-time and lap-time measurements are performed with black oxidized, counterbalanced chronograph and rattrapante hands.

PIAGET

PIAGET GOUVERNEUR CHRONOGRAPH – REF. G0A37112

The Piaget Gouverneur Chronograph houses the ultra-thin, self-winding 882P caliber, which is comprised of 33 jewels, possesses a power reserve of 50 hours and oscillates at 28,800 vph. Exhibiting technical expertise alongside a dashing appearance, this model is part of Piaget's Black Tie watch collection. The 18K pink-gold case matches the pink-gold appliqué hour markers, round counters and hands to a tee. These functions, which include a flyback chronograph as well as a second time zone counter and date window, rest beautifully atop a silvered guilloché dial.

PIAGET

PIAGET POLO FORTYFIVE CHRONOGRAPH – REF. G0A37004

A modern take on the renowned Piaget Polo watch, this handsome and athletic timepiece has a diameter of 45mm, which alludes to the 45-minute duration of a polo match. Piaget shows its daring side, as the case of this model is made of materials that were previously unused by the house. Composed of titanium and steel godroons with a black ADLC treatment, the exterior of this watch provides the wearer with maximum comfort. The black dial is complemented by bright white appliqué hour and minute markers as well as luminescent hands that guarantee legibility. Aside from the triple date window at 12:00 and the second time zone counter at 9:00, this watch possesses a flyback chronograph complete with a 30-minute counter at 3:00.

RICHARD MILLE

RM 039 TOURBILLON FLYBACK AVIATION E6B

To explore the themes of aviation and aeronautics in greater depth, Richard Mille has created a caliber designed for flight navigation. This complete and technically sophisticated instrument is an ultra-modern interpretation of an essential device used by every established pilot. The watch displays much of the same information provided by the famous E6B flight computer. This round slide rule, incorporated into the bidirectional rotating bezel, can be used to read off and calculate fuel burn, flight times, ground speed, density altitude or wind correction, as well as the fast conversion of units of measurement. The RM 039 tourbillon caliber is also equipped with a UTC hand, a countdown mode, an oversize date at 12:00 and a function selector. The movement drives a flyback chronograph featuring an exclusive design.

RICHARD MILLE

RM 50-01 TOURBILLON CHRONOGRAPH G-SENSOR LOTUS F1 TEAM ROMAIN GROSJEAN

Richard Mille Watches has been the official timekeeper of the Lotus F1 Team since 2013, and this fruitful collaboration is the first tourbillon watch dedicated to team member Romain Grosjean. RM 50-01 tourbillon uses a function indicator that shows whether the movement is in winding, neutral or hand-setting mode, with the addition of a mechanical G-force sensor, a mechanism able to transcribe the range of G's endured by the driver during deceleration phases. In total, the creation of this ultimate racing tool required more than 500 components.

VACHERON CONSTANTIN

OVERSEAS CHRONOGRAPH – REF. 49150/B01A-9745

Beating at a frequency of 21,600 vph, the Overseas Chronograph's automatic-winding 1137 caliber provides this timepiece with a precise set of functionalities. At 3:00 and 9:00, respectively, the chronograph's 30-minute and 12-hour counters join the small seconds at 6:00 and large date at 12:00, to complete the backdrop of the frontal view. While the stainless steel case and blue dial give the watch a powerful allure, the adornment of the caseback with the two-masted Overseas ship medallion reminds the wearer of watchmaking's rich nautical heritage.

ZENITH

EL PRIMERO 410 – REF. 03.2091.410/01.C494

The silver-toned sunray dial of this self-winding column-wheel chronograph sets a vibrant stage for the El Primero 410 caliber's wealth of indications. Seamlessly integrated within the architecture, the day and month, at 10:00 and 2:00 respectively, are displayed via slender apertures just above two traditional subdials indicating the running seconds and chronograph's 30-minute counter. At 4:30, a date window plays off the symmetry of the dial with its off-centered positioning. A dual subdial at 6:00 combines the chronograph's 12-hour totalizer with a colorful moonphase that echoes the bright blue of the 390-component, 36,000 vph movement's high-precision chronograph seconds.

ZENITH

EL PRIMERO LIGHTWEIGHT – REF. 10.2260.400/69.R573

An exhibition of state-of-the-art engineering, the El Primero Lightweight unites the brand's emblematic three-tone subdials with a decisively modern openworked architecture. Housed in a lightweight 45mm carbon case, the self-winding El Primero 400B caliber boasts a number of titanium components for an exceptional degree of performance. The high-tech spectacle uses silicon, notable for its anti-magnetic qualities, in both the lever and escape wheel. The central chronograph seconds and three counters are accompanied by the running seconds and chronograph's 30-minute and 12-hour totalizers. The innovative display at 6:00 uses a red accent that clearly identifies the date's appropriate numeral from a rotating skeletonized ring. This model won the "Sports Watch" award at the GPHG.

ZENITH

EL PRIMERO STRATOS SPINDRIFT – REF. 86.2060.4061/21.R573

Combining micro-mechanical elegance with a sporty personality, this 18K rose-gold and titanium timepiece honors its boat-racing inspiration while delighting its wearer by revealing its regulating organ on the dial side. The robust exterior, characterized by a rotating bezel and bold black pushbuttons, makes for a graceful contrast with the exhibition of the El Primero 4061 caliber's escapement at 10:00. Enhanced with silicon components, the self-winding 36,000 vph movement complements its animation of the hours and minutes with a high-precision chronograph comprised of a luminescent seconds hand and counters at 3:00 and 6:00. Running seconds are displayed at 9:00 within the colorful presentation of the caliber's inner workings.

Alarms

LONDON
AZORES
S.GEORGIA
RIO
CARACAS
NEW YORK
MEXICO
DENVER
L.ANGELES
ANCHORAGE
HAWAII
ON/OFF
FULL
ALARM
RESERVE
SWISS

Zenith
Pilot Doublematic

Alarms

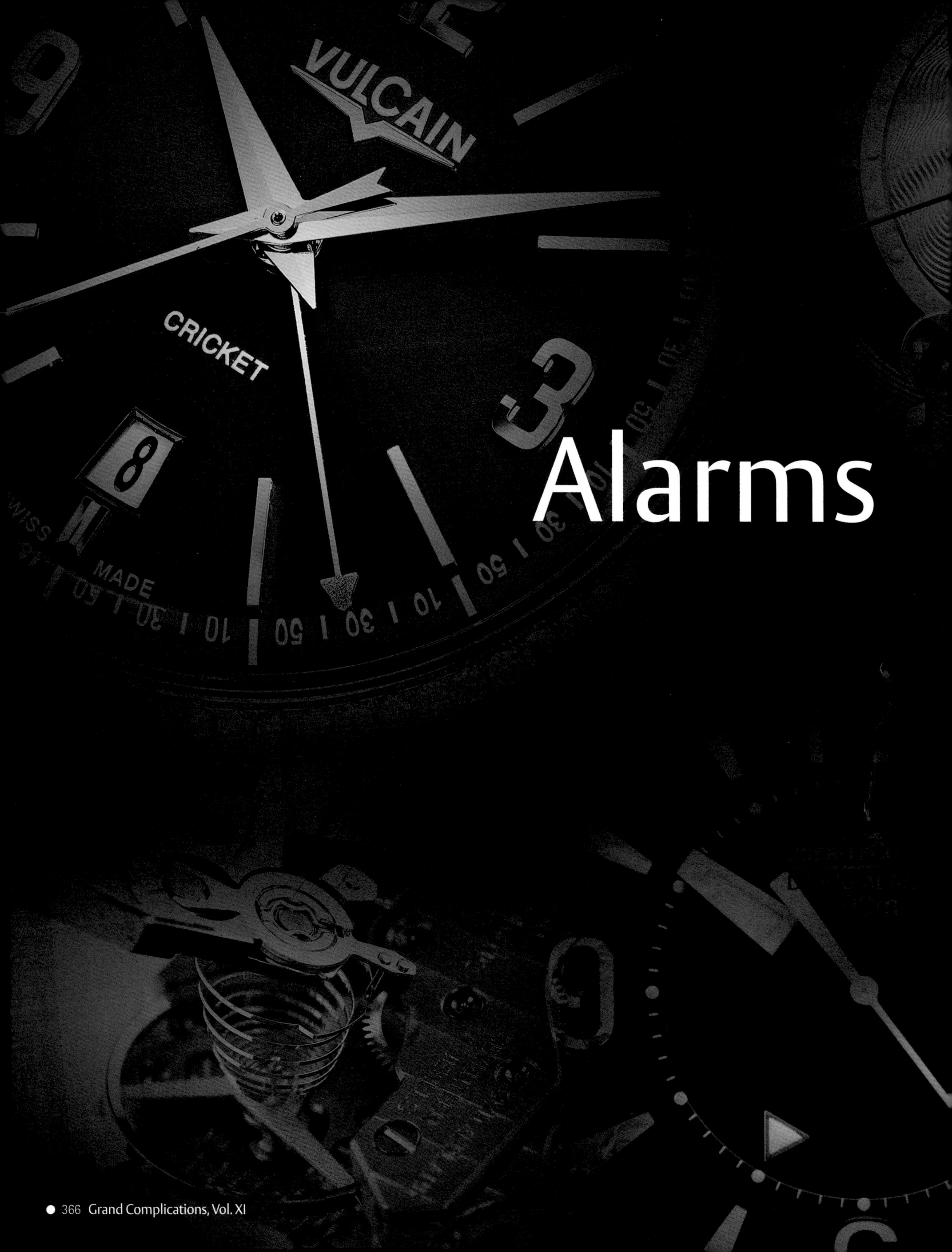

The world of horological complications goes through cycles of fashion like anything else; functions become fads, fall out of favor and make triumphant comebacks. Superstars in the 1950s and '60s, mechanical alarm watches then faded from the scene almost entirely, following the emergence of quartz watches and clock radio alarms. However, since the turn of the millennium, alarm watches have returned in full force, whether as reinterpretations of classic models or as new creations. To hold out against cellular phones and other modern conveniences, alarm watches have had to reinvent themselves—developing new displays, optimizing setting systems to enable users to set their alarms to the nearest minute or associating this functio n with other useful complications, such as travel or diving watches. New striking modes have appeared as well, often offering truly musical alarms.

THE ALARM WATCH IS UNDOUBTEDLY THE OLDEST HOROLOGICAL COMPLICATION. In fact, it seems that in the 13th century, Western monasteries were equipped with audible mechanisms intended to awaken monks so they could say morning prayers at the prescribed hour.

One should also note that the word "watch," which comes from the same root as "wake," first referred to "a clock that wakes one up," before being applied to any watch, no matter what functions it provided.

In the 15th century, table clocks were often equipped with an alarm mechanism that sounded on a small bell. This type of instrument would enjoy such high consideration that in 1601, the rules of the Corporation des Horlogers de Genève would demand that whoever wished to earn the certification of "maître" or master-watchmaker must produce a small clock with a morning alarm. The 16th century saw the appearance of pocket alarm watches with openworked casebacks to enhance the acoustics of the sound. In the 17th century, watchmakers also created oversized "carriage watches," with mechanisms to wake passengers in time to make their connections.

The first wristwatch alarms began to appear in 1914, but their sound was not yet strong enough to wake a sound sleeper. For that, the world would have to wait until 1947, when Vulcain launched the first truly functional alarm wristwatch, powered by the famously piercing Cricket caliber.

In the 1950s and '60s, the mechanical alarm watch experienced its golden age. Many watch brands were interested in this newly popular domain, developing many technical innovations, such as the first watches with a volume option (loud/soft) or with self-winding movements.

However, in the 1970s, the tidal wave of quartz movements and radio alarm clocks seemed to sign the death warrant for mechanical alarm watches.

The renaissance of mechanical watchmaking, at the end of the 1980s, signaled the grand return of the wristwatch alarm. For several years, this grand classic of horology has been enjoying renewed popularity among some of the biggest names in the industry. Technical enhancements abound, as do variations on display systems and designs—all of which are clear signs that this useful function will be making itself heard for some time to come.

HOW DOES IT WORK?

The alarm watch mechanism comprises two parts: one for setting the time, the other to activate the alarm strike. As a rule, it uses its own barrel (energy storage), separate from the barrel that powers the movement.

Alarm wristwatches most often distinguish themselves by the presence of an additional crown, which allows the wearer to both reset the spring of the striking mechanism and set the alarm.

Certain models also boast a small display on the dial indicating if the alarm is engaged or not.

The sound is produced by a small hammer that strikes a piece of metal. This metal piece could be a membrane with a pin, a kind of bell set into the case middle, or a gong (a steel wire wrapped around the movement), as in minute repeaters. Gongs produce a less powerful sound than the other two solutions. The alarm sounds until the striking work barrel is completely wound down.

Rereleasing the Classics

The history of the alarm watch has its stars, its legends, its flagship models. Riding the wave of enthusiasm for all things vintage, several famous brands have rereleased certain classic pieces, adapting them to modern standards of size and technical sophistication.

Tudor, *Heritage Advisor*

Inspired by a 1957 model, the Heritage Advisor from Tudor is equipped with an alarm module developed by the brand. The ON/OFF window at 9 o'clock indicates if the alarm function is engaged and the alarm's power reserve is displayed on a disc at 3 o'clock.

Vulcain, *50s Presidents' Watch*

Tribute to the famous "Presidents' Watch," first released in 1947, the 50s Presidents' Watch from Vulcain combines a design inspired by a 1950s model with the modern performance of the automatic-winding alarm caliber Cricket V-21.

Jaeger-LeCoultre, *Master Memovox International*

Jaeger-LeCoultre celebrated the 1958 launch of the Memovox Worldtimer (also known as Memovox "Heures Internationales") by creating the Master Memovox International. The dial's central disc, which allows the wearer to set the alarm time, is adorned with the names of 24 cities, representing the 24 time zones.

Jaeger-LeCoultre, *Master Memovox*

In 1956, Jaeger-LeCoultre launched the first automatic watch with an alarm function. The manufacture paid homage to this iconic model by creating the Master Memovox, with caliber 956 integrating the latest technical developments of the manufacture.

Corum

Heritage Vintage Chargé d'Affaires

A reinterpretation of a model first launched in 1956 in the first Corum collection, the Heritage Vintage Chargé d'Affaires limited series incorporates the same alarm movement as the original and rings for a full 12 seconds.

Variations on a Theme

In its most classic version, the alarm watch stands out by its small additional hand or the central disc for regulating the alarm. But the choice is ever-expanding, with multiple variations on the display mode or the style—whose openworked dials reveal the mechanisms within. On the technical side, the alarm increasingly accompanies other useful complications, such as the chronograph or the calendar.

Vulcain, *Golden Heart*

Due to its partially skeletonized dial, Vulcain's Golden Heart allows the wearer to admire the complex architecture of the manufacture's Cricket movement, with its anthracite coating and carefully executed finishings.

Fortis

B-43 Flieger Chronograph Alarm GMT

Created in 2012 for the 100th anniversary of Fortis, the B-43 Flieger Chronograph Alarm GMT boasts the first automatic chronograph movement with integrated alarm function—completed by a second time zone.

Alfex, *Pazzola Giant Alarm*

In 2009, the brand from Tessin, Switzerland, Alfex, celebrated its 60th anniversary by presenting the Pazzola Giant Alarm. One technical and aesthetic innovation: setting the alarm is done via a disc placed under the dial, positioned in relation to a red marker at 6 o'clock.

Jaeger-LeCoultre, *Master Grand Réveil*

A rare, even unique, example of an alarm watch with perpetual calendar, the Master Grand Réveil from Jaeger-LeCoultre can be set to either chime or vibrate. The alarm mechanism includes a barrel independent from the principal movement, as well as a hammer that makes a gong suspended in the caseback vibrate.

Oris, *Artelier Alarm*

Oris—known for its small alarm clocks in the 1940s—now offers an elegant, classic wristwatch named Artelier Alarm, with a central display separate from the alarm time and a chime specially developed by the brand.

Jaeger-LeCoultre, *AMVOX1 R-Alarm*

Born from the partnership between Jaeger-LeCoultre and Aston Martin, the AMVOX1 R-Alarm stands out for its sporty aesthetic that recalls automobile gauges. The alarm is set with a small hand on a minute track at the center of the dial.

Jaeger-LeCoultre Manufacture

Alarms one Can Set in Advance

On a traditional alarm watch, the alarm can only be set 12 hours in advance, i.e. in the evening for the next morning. For increased practicality, several brands have developed mechanisms that can set the alarm 24 hours ahead of time—or even a month in advance. Certain systems also set the alarm to the minute.

Harry Winston, *Project Z6 Black Edition*

Harry Winston has put its name to a highly original reinterpretation of the alarm watch in its limited edition Project Z6 Black Edition. The small off-center dial at 4 o'clock allows the wearer to set the hour and minute of the alarm up to 24 hours in advance, using two discs and a day/night indicator. The rectangular gong, cast in one piece, guarantees optimal acoustics.

Ulysse Nardin, *Sonata Streamline*

On the Sonata Streamline from Ulysse Nardin, the user sets the alarm using an hour/minute counter at 1:30. The 24-hour countdown display at 10:30 allows the alarm to be set the day before with no risk of confusion. The date, alarm and countdown automatically adjust to the selection of a new time zone.

Glashütte Original, *Senator Diary*

The German brand Glashütte Original, a branch of the Swatch Group, innovated with its presentation of the Senator Diary, the first alarm watch programmable for up to 30 days in advance. The crown at 10 o'clock and the pushbutton at 8 o'clock allow the wearer to determine the date and time of the alarm to within a quarter of an hour.

Musical Alarms

Moving away from more strident sounds, certain brands prefer to charm the ear by equipping their alarm watches with "cathedral" chimes worthy of the most beautiful minute repeaters, or even the kind of precious mechanisms that play entire tunes, as do music boxes. These dulcet tones soften the harsh reality of waking up—or remind the wearer with a few notes of an important meeting during the day.

Patek Philippe
Grandmaster Chime
Created for the 175th anniversary of Patek Philippe (1839-2014), the Grandmaster Chime presents, among its 20 complications, an alarm that sounds the time with the same striking sequence as the minute repeater—a world first. The alarm time can be set in 15-minute increments, but it always sounds two minutes before the full quarter-hour, to ensure that the strike can still be heard even at times that are signaled by a minimum of strikes, such as 1 o'clock.

Vulcain, *Imperial Gong*
The first wristwatch with an alarm and a tourbillon, Vulcain's Imperial Gong—launched in 2005—also stands out for its cathedral chime, worthy of the most beautiful minute repeaters.

Pierre DeRoche, *TNT RendezVous*
The TNT RendezVous from Pierre DeRoche is not designed to wake a slumberer. It prefers to discreetly signal a moment chosen in advance by a series of double chimes. Its two hammers strike two "cathedral" gongs partially visible through the dial.

Boegli, *Grand Orchestre*
Equipped with a tiny 17-note music box, the Grand Orchestre model by Boegli offers a chance to wake up gently to the sound of Mozart's "Magic Flute," Vivaldi's "Four Seasons," or Beethoven's "Für Elise."

Breguet, *Réveil Musical*
Masterpiece of haute horology, Breguet's Réveil Musical possesses a patented mechanism inspired by music boxes, with the cylinder replaced by a disc with pins. During the melody, which lasts about 20 seconds, the dial completes a full rotation. The automatic caliber is equipped with a silicon escapement.

Travel Alarms

What could be more useful than an alarm while one is traveling? Inspired by this rhetorical question, watchmakers are increasingly combining the alarm function with a second time zone or "world time" display (see chapter on Multiple time zones and GMTs). The most sophisticated models stand out due to their alarm mechanism that is always linked to the local time.

Blancpain, *Léman GMT Alarm*

On the Léman GMT Alarm from Blancpain, the alarm is indexed to the local time display, indicated by the large hour and minute hands. Another practical innovation: the automatic-winding mechanism allows the wearer to charge the power reserve and the alarm simultaneously.

Panerai, *Radiomir GMT Alarm*

The Radiomir GMT Alarm from Panerai combines a digital display of a second time zone at 6 o'clock with an alarm indicated by a central hand. The two functions are controlled by a crown located on the right side of the case.

Vulcain, *Aviator GMT Pilot DLC*

The Aviator GMT Pilot DLC model by Vulcain combines a hand-wound alarm caliber with a world-time display. The inner bezel ring indicating the 24 reference cities is adjusted via the crown at 4 o'clock.

Jaeger-LeCoultre

Master Compressor Extreme W-Alarm Aston Martin, Limited Edition

The Master Compressor Extreme W-Alarm Aston Martin from Jaeger-LeCoultre, with a world time function, presents several patented innovations in its alarm mechanism, which is equipped with a gong wrapped around the movement. The time of the alarm can be precisely selected using a double disc with hours and minutes in a window at 9 o'clock.

Breguet, *Le Réveil du Tsar*

The Réveil du Tsar, from Breguet, possesses an alarm mechanism indexed to the local time, with an alarm engagement indicator in a small window at 12 o'clock. The hours/minutes subdial, at 3 o'clock, allows the wearer to set the alarm with to-the-minute precision.

Yeslam, *Ti2 Réveil*

Equipped with a movement from La Joux-Perret, the Ti2 Réveil from Yeslam contains within its square titanium case an alarm with a central hand, an aperture for the date and a 24-hour time zone with a digital display.

Underwater Sounds

Many diving watches are equipped with an alarm, just as some alarm watches are getting their sea legs. This might seem like an odd alliance, as the ocean depths are not ideal for spending the night, or even for a small nap. But here the alarm is used to remind the wearer of a precise moment, notably the end of the dive time—and it doesn't hurt that sound carries much better underwater than in open air.

Jaeger-LeCoultre
Memovox Tribute to Deep Sea

In homage to Memovox Deep Sea, the first Jaeger-LeCoultre diving watch, released in 1959, the Memovox Tribute to Deep Sea is powered by the automatic Memovox 956 caliber, with hours, minutes, seconds and alarm. The caseback reprises the timepiece's original design of a frogman surrounded by bubbles.

Jaeger-LeCoultre
Master Compressor Diving Alarm Navy SEALs

Jaeger-LeCoultre celebrated its partnership with the elite combat swimmers through the 2009 launch of the Master Compressor Diving Alarm Navy SEALs, a limited edition with a grade-5 titanium case, a ceramic bezel and water resistance to 300m.

Vulcain, *Diver X-Treme Automatic*

Vulcain has developed several watches that combine an alarm with underwater capability, such as the Diver X-Treme Automatic, with a case that is water resistant to 100m, audible underwater strike, rotating 60-minute bezel to calculate the dive time and an ultra-readable dial highlighted with luminescent indexes.

Vulcain, *Nautical Heritage*

In a rerelease of a famous 1961 model, the Nautical Heritage from Vulcain—water resistant to 300m—stands out for its alarm that is audible underwater, its central rotating dial displaying decompression stages and its triple-backed case, which serves as a resonance chamber for the chime.

Breguet, *Marine Royale*

Breguet has released its sporty Marine Royale collection in an alarm model water resistant to 300m, with a modern aesthetic that combines rubber and rose gold.

BREGUET

CLASSIQUE LA MUSICALE – REF. 7800BA.11.9YV

A masterful fusion of innovation, sublime craftsmanship, and design ingenuity, this self-winding 48mm timepiece in 18K yellow gold conducts a brilliant visual accompaniment to its pioneering musical score. Equipped with the first ever magnetic strike governor, a patented system of isolated magnets used to eliminate background noise, the melodic tune, pre-set to a desired time or activated via a pushbutton at 10:00, is joined by the superb and complete rotation of the hand-guilloché dial. Powered by the 55-jewel 0900 caliber with silicon escape wheel and Breguet balance and spring, this horological work of art is finished with a caseband intricately engraved with a musical stave.

VULCAIN

50S PRESIDENT'S WATCH "VULCAIN FOR HERBIE HANCOCK" – REF. 10651.301L

Manifesting the free-spirited character of jazz music and its fondness for exploring untapped territories, the "Herbie Hancock" 50s Presidents' Watch by Vulcain features a sunburst dial in a bright shade of blue. Complete with the hand-wound Cricket V-16 movement, the watch displays hours, minutes, seconds, date and alarm. Measuring 42mm in diameter, the case is engraved on the back with the name of the celebrated jazz musician. This stainless steel model is available in a limited edition of 250 pieces while the 18K pink-gold version is issued in a series of 50.

VULCAIN

HERITAGE PRESIDENTS' WATCH – REF. 100653.290L

Vulcain has paid tribute to its legendary Presidents' watch with this model, a faithful replica, which features the latest technological innovations. Complete with the famous Cricket V-10 caliber, which oscillates at 18,000 vph, it provides a power reserve of 42 hours. The movement features twin barrels, one to supply the energy required by the hour, minute and seconds functions, the other to power the alarm function and its 20-second chime. Composed of 157 parts, it features, hours, minutes, central and date functions alongside the alarm that ensured Vulcain's fame over sixty years ago.

ZENITH

PILOT DOUBLEMATIC – REF. 03.2400.4046/21.C721

The Pilot Doublematic is a open invitation to travel. This chronograph features an alarm function controlled by the pushbutton and the crown at 8:00. The Pilot Doublematic also defies time differences, thanks to two discs that simultaneously display the time in different parts of the world. Beating at 36,000 vph and endowed with a 50-hour power reserve, the automatic-winding El Primero Calibre 4046 may be admired through an openworked oscillating weight that is itself visible through the sapphire caseback.

Multiple Time Zones & World Time Indications

de GRISOGONO

Instrumento N° Uno

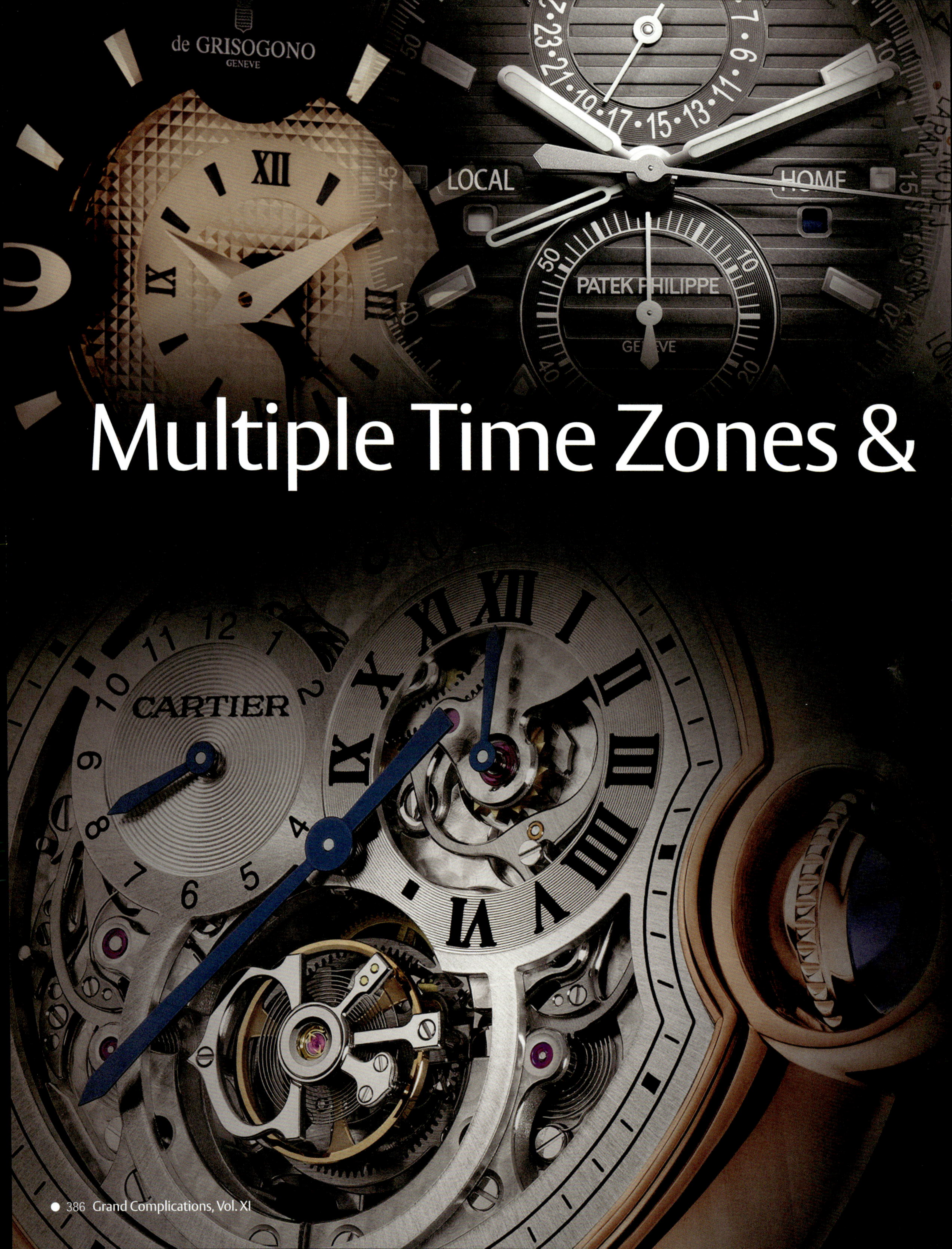

Multiple Time Zones &

World Time Indications

Global watches for a global community! The popularity of international travel and communications across the planet has led to the unprecedented success of watches that simultaneously display two, three, four or even 24 time zones—or more! With pointers, discs, bezels, star maps or small three-dimensional globes, brands are developing a wide palette of technical and aesthetic solutions to create the most practical—or the most attractive—time displays. Watchmakers are also applying themselves to the challenge of devising the most intuitive, user-friendly possible methods of setting or correcting the time. From watches equipped with dual hour hands to the famous "world time" models indicating the time in all time zones at once, today's globe-trotters and businesspeople have a wide array of choices to accompany them on their next trip.

BEFORE THE DIVISION OF THE GLOBE INTO 24 TIME ZONES, each region lived at its own pace, and travelers had to adjust their watches constantly as they moved from one place to another.

There were a multitude of local times, based, more or less, on longitude: more than 70 in North America and around 30 in Europe. The increase in long-distance travel, and especially the birth of the railroad, generated a need to create a unified system of time. In 1847, the British Railway Clearing House recommended that all companies adopt the time at the Greenwich Observatory (Greenwich Mean Time, or GMT), which already served as a reference for navigators. In 1884, at the International Meridian Conference in Washington, DC, the decision was taken to divide the planet into 24 time zones, each one covering 15° of longitude (one hour), and to assign longitude 0 to the Greenwich meridian. This system was rapidly adopted by most of the countries in the world—except France, which did not come around until 1911.

The first dual time zone watches, sometimes called "captain's watches," were equipped with two movements, more or less coordinated, that displayed the time zones on two different dials. Later, watchmakers perfected this system by using just one movement, equipped with a mechanism uncoupling the additional hour hand.

Though watches with a multitude of local times already existed as early as the 17th century, it was not until the 1930s that a Genevan watchmaker, Louis Cottier, invented the ingenious "world time" system, which allows the wearer to read the time in all time zones simultaneously.

GMT AND UTC

Greenwich Mean Time corresponds to the mean solar time at the meridian crossing the Greenwich Royal Observatory, near London. It was adopted at the end of the 18th century as a reference time zone for all others on the planet, and is based on the rotation of the Earth. It was replaced in 1972 by "Universal Coordinated Time," better known by its acronym UTC, which is based on International Atomic Time (with the acronym TAI, for its French name). Watchmakers commonly use the term GMT to designate a watch indicating a second time zone, even if that time zone is not Greenwich.

HOW DOES IT WORK?

Multiple time zone watches offer such a variety of technical approaches and methods of use, that it is impossible to describe in detail all the particularities of their functioning.

Generally, they possess an additional display for the hour that is powered by the main system, but can be temporarily disengaged in order to set it to local time—without affecting the accuracy of the minutes and seconds. A pushbutton or the crown serves as a tool for setting the second time zone.

In order not to awaken a friend or a business partner who might be sound asleep, travelers often need to know if it is day or night in their home region. This is why some watches display home time on a 24-hour scale. Others use a 12-hour display, but supplement this with a day/night indication (a small aperture or subdial that changes from white to black according to the time of day), a small additional 24-hour scale or an AM/PM display.

Many watches boast another very useful function when one is traveling: automatic linkage of the date to local time.

WORLD TIME

To continuously display the time in all corners of the globe, world time watches possess—in addition to the central hands—two mobile discs, one with 24 gradations, the other bearing the names of the reference cities. The 24-hour disc turns counterclockwise.

Double Hour-Indication Systems

The simplest system for simultaneously displaying two time zones is to add an hour hand with a distinctive design, either to the center of the dial, or—more rarely—on a small auxiliary dial. The second time zone might also distinguish itself with a retrograde or jumping display, or it might appear only when activated by the user. The most practical models equip their home time zone with a 24-hour scale or a day/night indicator.

Cartier, *Ballon Bleu de Cartier Flying Tourbillon 2nd Time Zone*

The Ballon Bleu de Cartier Flying Tourbillon 2nd Time Zone is inspired by regulator-type dials with a central minute hand and two jumping-hour totalizers at 10 o'clock and 2 o'clock respectively.

de GRISOGONO, *Fuso Quadrato*

On the Fuso Quadrato from de GRISOGONO, the hand for the additional time zone is hidden beneath a dial composed of 12 titanium shutters that opens on demand using a slide at 9 o'clock.

Maîtres du Temps
Chapter Three Reveal
The independent master-watchmakers Kari Voutilainen and Andreas Strehler designed a two-faced watch for the brand Maîtres du Temps. A pushbutton in the crown lowers two "panels" to reveal two additional indications on rollers: a second time zone and a day/night indicator.

Louis Vuitton, *Tambour éVolution Spin Time GMT*
The Tambour éVolution Spin Time GMT from Louis Vuitton indicates the hours from a second time zone using a dozen pivoting rods, each one topped by a small cube. At the turn of the hour, one of the cubes rotates to reveal an arrow pointing outward or inward, to the appropriate hour on a 24-hour scale. Meanwhile, the cube representing the preceding hour makes a quarter turn to present a blank side.

DeWitt, *Twenty-8-Eight Double Fuseau GMT2 Poetic*
On the Twenty-8-Eight Double Fuseau GMT2 Poetic from DeWitt, the second time zone is displayed by a small, red, sword-shaped, open-worked pointer, complemented by a day/night indicator at 12 o'clock that shows the sun or moon through an aperture in the dial.

Hermès, *Dressage L'heure masquée*

On the highly original Dressage L'heure masquée from Hermès, the hour hand is concealed behind the minute hand, appearing only with the push of a button incorporated into the crown. This same push allows the wearer to read the time in a second time zone through an aperture at 6 o'clock.

IWC

Ingenieur Dual Time Titanium

On the Ingenieur Dual Time Titanium by IWC, the main hands show the local time, while the hand with the white triangle provides a 24-hour dual-time display. Pushpieces serve to adjust the time in one-hour increments.

Ulysse Nardin, *Dual Time Manufacture*

The Dual Time Manufacture from de Grisogono stands out for its small round aperture at 9 o'clock that continuously displays home time on a 24-hour scale, while the pushbuttons at 8 and 10 o'clock allow the user to set the local time with the central hand, accompanied by automatic date adjustment.

Breitling
Chronomat 44 GMT

The Chronomat 44 GMT from Breitling houses an in-house movement whose dual time zone function is equipped with a simple, fast time-setting system via the crown. The rotating bezel displays a third time zone on a 24-hour scale.

Van Cleef & Arpels,
Pierre Arpels Heure d'ici & Heure ailleurs

On the Pierre Arpels Heure d'ici & Heure ailleurs model from Van Cleef & Arpels, the two hour hands are replaced by two jumping-hour apertures at 11 o'clock (local time) and 5 o'clock (home time), while the retrograde minute display runs along an arc from 5:30 to 10:30.

Rolex, *Oyster Perpetual GMT-Master II*

The Rolex Oyster Perpetual GMT-Master II is a classic dual time zone watch launched in 1955 that solves the problem of daytime and nighttime hours by means of a 24-hour display of the second time zone along with a two-tone red and blue Cerachrom ceramic bezel to indicate day and night.

Superimposable Hands

On most travel watches, the additional time zone hand is always visible, even when the wearer has no use for it. Some watchmakers have improved upon this "defect" by devising hour hands that can be superposed, or placed one on top of the other, and function in a perfectly synchronized manner. This is a means of offering two watches in one.

Jaeger-LeCoultre
Master Hometime
The Master Hometime from Jaeger-LeCoultre proposes a system of dual crown-activated superimposable hour hands. The skeletonized hand on a 24-hour scale maintains home time. The date is linked to the local time.

H. Moser & Cie, *Endeavor Dual Time*

On the self-winding Endeavor Dual Time by H. Moser & Cie, two superimposed hour hands are set via a patented crown-adjustment system. The white or black indicator at 12 o'clock distinguishes between day and night (AM or PM hours) in the wearer's home time zone.

Patek Philippe, *Nautilus Travel Time Chronograph Ref. 5990*

On the Nautilus Travel Time Chronograph Ref. 5990 from Patek Philippe, the pushbuttons allow the wearer to advance or push back the local time indicator in one-hour increments. When the dual time zone function is not being used, the two hour hands appear as one. This watch is also equipped with an automatic chronograph.

Jaquet Droz, *Grande Heure GMT*

Bearing just two superimposed compass-type hour hands—a nod to historical navigating instruments—the Grande Heure GMT by Jaquet Droz provides an intuitive 24-hour dual-time reading.

Double Hour/Minute Displays

Though the minutes generally remain the same from one time zone to the next, certain high-end watch brands prefer to include a complete display of the time in each region, with a second set of both hour and minute hands located in a small auxiliary dial. Another possibility involves indicating the two time zones on two separate dials, in watches featuring a resolutely distinctive design.

de GRISOGONO, *Meccanico dG*

de GRISOGONO's astonishing Meccanico dG combines an analog display for the first time zone and a digital display, driven by 23 cams, for the second time zone.

Breguet, *7067 Tradition GMT*

Equipped with an openworked architecture that reveals the principal components of the movement, the 7067 Tradition GMT from Breguet displays local time at 12 o'clock. A guilloché subdial at 8 o'clock displays home time, complete with day/night indicator.

Piaget, *Altiplano Double Jeu*

Piaget's Altiplano Double Jeu possesses two superimposed cases, each containing an ultra-flat mechanical movement. The dial of the lower case displays a second time zone on a 24-hour scale.

Oris
Greenwich Mean Time Limited Edition

In 2014, to commemorate the 130th anniversary of the 1884 division of the planet into 24 time zones, Oris created the Greenwich Mean Time Limited Edition, with a double hour/minute display. The central hand is equipped with a system to correct the time forwards or backwards, in one-hour increments, activated by two pushbuttons. The subdial at 3 o'clock has a day/night indicator.

Rolex, *Cellini Dual Time*

The new Cellini collection from Rolex, launched in 2014, notably includes a Dual Time model, whose subdial at 6 o'clock displays a second time zone with hours and minutes, as well as a day/night indication in the form of a miniature sun or moon visible through an aperture.

Jaeger-LeCoultre
Reverso Grande GMT

The Reverso Grande GMT from Jaeger-LeCoultre makes the most of its famous reversible dial by proposing two time zones on two dials that lie back to back. The dial on the reverse side includes an indicator for the number of hours' time difference between local time and Greenwich Mean Time or another reference point.

Adjustment by Half Time Zones, Quarter Time Zones, or to the Nearest Minute

On the vast majority of travel watches, the second time zone is adjustable in increments of one hour. But there are countries in which the deviation from UTC adopts finer divisions of half an hour or even a quarter of an hour. This is notably the case with Iran (UTC + 3h30), India (UTC + 5h30) Nepal (UTC + 5h45) and the middle of Australia (UTC + 9h30). A few watches take these particularities into account.

Arnold & Son, *DBG*

The DBG by Arnold & Son features two separate hour/minute displays, powered by their own barrels, going trains, escapements and balances, and independently adjustable to the nearest minute. The subdial at 12 o'clock displays the difference between the two time zones.

Glashütte Original,
Grande Cosmopolite Tourbillon

The Grande Cosmopolite Tourbillon from Glashütte Original can display two hour/minute time zones from among a selection of 37 locations (including those that vary by 30 or 45 minutes), with a distinction between summer and winter times. Thanks to its perpetual calendar, it also takes into account the exact monthly and leap-year cycle for these calculations.

Blancpain
Villeret 8 Day Half-Timezone

The Villeret 8 Day Half-Timezone from Blancpain allows the wearer to set the second time zone (which includes hours and minutes) in increments of half an hour using a crown equipped with an integrated pushbutton.

Parmigiani, *Tonda Hemispheres*

Parmigiani's Tonda Hemispheres is equipped with a small auxiliary dial with hours and minutes, located at 12 o'clock and adjustable to the minute using a second crown at 2:30. Both time zones have day/night indicators. This model is notably offered with a dial featuring SuperLumiNova-enhanced counters.

Jaeger-LeCoultre
Duomètre Unique Travel Time

The Duomètre Unique Travel Time from Jaeger-LeCoultre displays a second time zone on a subdial at 9 o'clock with a jumping hour and minute hand adjustable to the nearest minute. The globe at 6 o'clock enables an intuitive reading of the time around the world.

Systems with Discs or Rotating Bezels

Watches with multiple time zones might operate using systems based on discs or rotating bezels—either in lieu of a pointer-type system or in conjunction with it. Watchmakers generally take this opportunity to introduce complementary indications, such as the names or abbreviations for certain cities, representing precise time zones.

Cartier, *Tortue XXL Multiple Time Zone*

On the Tortue XXL Multiple Time Zone watch by Cartier, a pushpiece and a lateral city disc serve to select a second time zone while taking account of "summer" or daylight saving time. The local time is displayed by the main hands, while the time in the user's home zone appears through an aperture with a day/night hand incorporating symbols of the sun and moon.

Jacob & Co., *Time Traveler Epic SF 24*

The Time Traveler Epic SF 24 model by Jacob & Co. shows the time in one of the world's 24 time zones through a large aperture at 12 o'clock, in a style inspired by airport display boards.

DeWitt, *Academia Hora Mundi*

Adorned with the iconic DeWitt ratcheted bezel, the Academia Hora Mundi serves to select the time zone (city) via a double aperture controlled by a dedicated pushpiece.

Hysek, *Verdict Automatic Triple Time Zone*

On the Verdict Automatic Triple Time Zone watch by Hysek, the second and third time zones are indicated by retrograde displays in the dial center, coordinated by two apertures displaying the names of cities respectively situated to the east and west of the Greenwich meridian.

Vogard, *Datezoner*

The Datezoner from Vogard stands out for its patented system that allows the wearer to change the time zone and the date simultaneously, via the bezel. One needs only to pull the security lever and rotate the bezel to select the desired city.

A. Lange & Söhne, *Lange 1 Time Zone*

A. Lange & Söhne's Lange 1 Time Zone possesses two auxiliary hour/minute dials with day/night indicators. The wearer adjusts the second time zone with a pushbutton that turns the rotating city ring.

Audemars Piguet, *Tourbillon Royal Oak Concept GMT*

On the Tourbillon Royal Oak Concept GMT from Audemars Piguet, the second time zone is displayed with the help of two stacked discs. The top disc, which completes a rotation in 12 hours, bears the numbers 1-12. The bottom disc, which completes a rotation in 24 hours, is divided into a white section for daytime and a black section for nighttime.

Corum, *Admiral's Cup Challenge 48 Day & Night*

The Admiral's Cup Challenge 48 Day & Night from Corum is adorned with a map of the world that bears a transparent rotating disc. This system can perform two functions: it can display a second time using a small white indicator, or it can indicate which side of the Earth is illuminated by the sun and which is in darkness.

Nomos, *Tangomat GMT*

On the Tangomat GMT from the German brand Nomos, the central time and the name of the reference city (indicated by abbreviated airport codes through an aperture at 9 o'clock) may be simultaneously adjusted by pressing a pushpiece at 2 o'clock, while a 24-hour second time-zone display appears through an aperture at 3 o'clock.

ochs und junior, *Due Ore*

The epitome of ingenuity and restrained aesthetics, the Due Ore watch by ochs und junior, developed by Ludwig Oechslin, displays a second time zone through apertures by means of a crown-adjustable rotating disc.

Three-Dimensional Planets

After the appearance of three-dimensional moons (see chapter on Moonphases), a new type of travel watch emerged, one that displayed small terrestrial globes rotating on the dial. These innovative, spectacular creations are most often produced by young, pioneering brands or independent watchmakers.

Hysek, *Colosso*

On the Colosso from Hysek, the terrestrial sphere, rotating over 24 hours, can be linked to either local time (on the central hands) or to a second time zone. The latter is indicated by a double linear retrograde display, with an aperture that specifies the reference city.

Magellan Watch, *Magellan 1521 NH*

Using a third hand tipped with a small solar disc, the Magellan 1521 NH displays either the position of the sun in relation to the Earth (longitude where it is noon), local time on a 24-hour scale or a second time zone, also on a 24-hour scale.

Greubel Forsey, *GMT*

The titanium globe adorning Greubel Forsey's GMT allows the wearer to take in at a glance all the Earth's time zones—complemented by a day/night indicator. For a more precise reading, the watch also displays a second time zone at 10 o'clock and a world time indication on the back of the platinum case.

VicenTerra, *GMT-3/T2*

On the VicenTerra GMT-3/T2, the small rotating sphere placed at 5 o'clock shows the longitude where the sun is at its height. This unusual system is completed by a second time zone at 7 o'clock, a day/night indication (sun and moon) at 12 o'clock and a retrograde date display at 2 o'clock.

World Times

Queen of travel watches, the world-time complication allows the wearer to see what time it is in all 24 time zones. To this end, they are usually equipped with a mobile ring that bears the names of 24 reference cities. A few watchmakers have chosen to reinterpret this classic model by varying the configuration of the dials, taking partial time zones into account or displaying specific time slots, such as the opening of the world's stock markets. The difference also finds scope for expression in the adjusting systems, whose practicality varies with the model.

Patek Philippe, *World Time Ref. 5131*

Heir to a long tradition, the World Time from Patek Philippe is distinguished by its dial in cloisonné enamel. To change the main time zone, the wearer uses the sole pushbutton, at 10 o'clock—the mechanism then simultaneously adjusts the city disc, the 24-hour disc and the hour hand.

Vacheron Constantin
Patrimony Traditionnelle World Time

The Patrimony Traditionnelle World Time by Vacheron Constantin simultaneously displays no fewer than 37 time zones—including those that differ from their neighbors by a half-hour or a quarter-hour. This exclusive piece is patented.

Montblanc
TimeWalker World-Time Hemispheres

The TimeWalker World-Time Hemispheres watch by Montblanc is available in two versions: one for each hemisphere. Its dial displays a map of the Earth seen from the North or South Poles, with a corresponding choice of cities.

Breitling, *Transocean Unitime Pilot*

The world-time chronograph Transocean Unitime Pilot associates an in-house Breitling movement with a technical black/white look inspired by aviator watches. One need only rotate the crown forwards or backwards to adjust all the indications with a single gesture, complemented by automatic date adjustment linked to local time.

Antoine Preziuso, *Transworld*

Transworld, from Antoine Preziuso, offers a very attractive reinterpretation of world time, with a mobile world map in the center of the dial, depicting the Earth as seen from the South Pole and making a complete rotation over 24 hours. The display is as original as it is poetic.

Baume & Mercier
Capeland Worldtimer

Baume & Mercier has enriched its Capeland collection with a Worldtimer model, a steel or red-gold watch featuring a world-time function powered by a self-winding manufacture movement visible through a transparent caseback.

Tissot, *Heritage Navigator*

In 2013, Tissot celebrated its 160th anniversary by re-issuing a 1953 model. Equipped with a chronometer-certified movement, this Heritage Navigator watch displays the time in all 24 time zones by means of the city names appearing on the dial.

Louis Vuitton, *Escale Worldtime*

On its Escale Worldtime model, Louis Vuitton has reinterpreted the world-time concept with a system devoid of hands, directed entirely from the crown. Three mobile discs show the hour and minute of the principal time zone below the name of the selected city at 12 o'clock. The highly colorful dial décor is inspired by motifs used to decorate Louis Vuitton suitcases.

Andersen Genève, *Communication 750**

On the Communication 750* by Andersen Genève, the central blue-gold dial completes one revolution per day while indicating the 24 time zones opposite the names of 30 major cities on five continents.

Girard-Perregaux, *Traveller ww.tc*

The Traveller ww.tc (world-wide time control) model by Girard-Perregaux teams an automatic chronograph with a world-time display. Its manufacture movement is housed in a titanium case with black DLC coating. The dial design evokes the criss-cross pattern of Earth's lines of latitude and longitude.

A. LANGE & SÖHNE

LANGE 1 TIME ZONE LUMINOUS – REF. 116.039

Operating on the solid silver dial of the 18K white-gold Lange 1 Time Zone, the luminous rhodiumed-gold hands and appliqués on this timepiece guide their owner even in the darkness of night. Conveniently adjusted via a pushbutton, the peripherally rotating cities ring offers a cosmopolitan and clear view of the applicable time zone's chosen reference location. Furthermore, this manual-winding timepiece is armed with a 72-hour power reserve, and boasts an ingenious synchronization mechanism that permits the seamless swapping of the indicated time between the two subdials, both of which possess an independent day/night indicator.

BELL & ROSS

BR126 SPORT HERITAGE GMT & FLYBACK

The Vintage Sport Heritage range pays tribute to watches worn by civil aviation pilots in the 1960s. This limited edition BR126 Sport Heritage GMT & Flyback (500 pieces) features both a second time zone and a chronograph function on its sleek black dial, as well as a tachometer scale. The second time display, on a 24-hour scale to differentiate between AM and PM, occupies a subdial at 9:00, with another subdial at 3:00 for running seconds. An aperture at 4:30 reveals the date. The 43mm case in polished steel frames an immovable black bezel with aluminum ring. The automatic-winding movement is visible through the sapphire crystal screw-down caseback.

de GRISOGONO

DOPPIO TRE – REF. N04

With three zones on display, the Doppio Tre uses both front and back for maximum functionality. All three time displays are powered by a single movement, which boasts 25 jewels and more than 300 components. Crafted in polished 18K rose gold, the reversible case bears three crowns; the crown at 12:00 controls the time on the flip side of the case. The black guilloché dial features rose-gold Arabic numerals, hour markers and hands. A date at 7:00 adds even more practicality to this extremely useful timepiece, which is mounted on a black alligator leather strap with an 18K rose-gold butterfly buckle.

de GRISOGONO

INSTRUMENTINO BLUE SAPPHIRES – REF. S24S AT

Instantly recognizable for its gently curved silhouette, the Instrumentino, launched in 2001, features perfect proportions, size and volume, making it supremely comfortable to wear. Powered by the DF 26-91 automatic movement, this watch displays a unique dual time zone function, with the second time zone corrector at 4:00. The polished 18K white-gold case is set with 88 baguette-cut blue sapphires (approximately 9.81 carats), and the crown is set with de Grisogono's characteristic black diamond. The pastel blue dial is set with a ring of 82 white diamonds, and the ensemble is mounted upon a pastel blue galuchat strap, another signature material of the brand.

de GRISOGONO

INSTRUMENTINO WHITE DIAMONDS – REF. S21 AT

The Instrumentino is 15% smaller than the Instrumento, making it the perfect size for a women's watch. This version, covered in white diamonds, is powered by the automatic DF 26-91 caliber, which possesses a 39-hour power reserve and oscillates at 28,800 vph. The movement's oscillating weight is engraved with the word "Instrumentino." The rose-gold case is fully set with 405 white diamonds (approximately 6.48 carats). The dial is fully set with 312 white diamonds, and the white-gold de Grisogono folding clasp linking the ends of the white galuchat strap is set with 88 diamonds.

de GRISOGONO

INSTRUMENTO N° UNO – REF. DF N11

This timepiece's black lacquered dial is enriched by the warmth of its case in 18K pink gold and echoed by a crown adorned with a black-diamond cabochon. On the dial, gradually enlarged 18K pink-gold Arabic numerals lead the eye from the primary time display on the upper half to the self-winding watch's secondary time zone directly beneath it. A white date aperture below the numeral "8" completes a creative and balanced geometry, and the movement may be admired through a sapphire crystal caseback.

de GRISOGONO

INSTRUMENTO Nº UNO – REF. DF S30

After two years of painstaking design, de Grisogono introduced its very first wristwatch in 2000. de Grisogono's brand philosophy is clearly conveyed through a combination of watchmaking tradition and unparalleled aesthetics, featuring de Grisogono creative twists. Powered by the automatic 197-component DF 11-89 caliber, the oscillating movement of which features the engraved words "Instrumento Nº Uno," this model is set with brown diamonds on the case and lugs, as well as white diamonds on the dial, with 12 hour markers in black diamonds and a signature de Grisogono black diamond on the crown.

F.P. JOURNE

CHRONOMETRE A RESONANCE

The iconic Chronomètre à Résonance of the F.P. Journe High Horology collection represents one of the wildest challenges ever in the field of the mechanical watch. Utilizing the phenomenon of natural resonance, this exceptional mechanism with two perfectly synchronized beating mechanical hearts revolutionized the classic standards of watchmaking by offering unequalled precision for a wristwatch. The case is available in platinum or red gold, with a white-gold and silver dial, and an 18K rose-gold movement.

FREDERIQUE CONSTANT

WORLDTIMER MANUFACTURE – REF. FC-718MC4H4

Complete with perlage and circular Côtes de Genève decoration on the movement, this model's gleaming silver dial features an intricate guilloché design at the center, surrounded by striking, hand-polished black oxidized hands and Roman numeral hour markers. In addition to the extra-large date counter at 6:00, a 24-hour disc with day-night indicators surrounds the Roman numerals. Surrounding the outer ring of the watch face, the Worldtimer feature shows the time in 24 of the world's greatest cities simultaneously. This model is complete with a glossy black alligator leather strap with a rose-gold pin tongue buckle.

GIRARD-PERREGAUX

TRAVELLER JOHN HARRISON – REF. 49655-52-133-BBBA

This model is perfect for the watch enthusiast on the go, as it pays tribute to John Harrison, the renowned English clockmaker who is known as the inventor of the marine chronometer. Housed in a 44mm pink-gold case, this model displays the large date at 12 with a second time zone at 4:30. The second time zone counter overlaps with the small seconds counter at 7:30, which also acts as a moonphase indication. The Traveller John Harrison is powered by the self-wound GP03300-0093 caliber, which bears 35 jewels, oscillates at 28,800 vph and has a power reserve of 46 hours.

GUY ELLIA

JUMBO HEURE UNIVERSELLE

One of the latest creations by designer Guy Ellia is the Jumbo Heure Universelle with a movement created by Frédéric Piguet. Caliber PGE 1150 features a 72-hour power reserve, a blue sapphire disc, and Côtes de Genève-finished bridges with rhodium plating. The day/night indicator, the grand date and 24 time-zone indications define its singular functions. The Jumbo Heure Universelle bears an exceptional 50mm size that is rarely seen in the world of luxury watches. Mounted on an alligator strap with a folding buckle, this model is presented in black gold and is also available in white or pink gold.

IWC

PORTOFINO MIDSIZE AUTOMATIC DAY & NIGHT – REF. IW459102

Thanks to its elegant design with a diamond-set bezel and white mother-of-pearl dial, the Portofino Midsize Automatic Day & Night ensures that its wearer cuts a fine figure no matter what the time. The watch features a sumptuous red-gold case, combined with a black alligator leather strap made by Santoni. Supplying the power is the 35700-calibre mechanical movement with a 42-hour power reserve. While the hour, minute and seconds hands display local time, the blue hand on the inner 24-hour ring shows the time at a second location of the wearer's choice. To make reading the time in the second location easier, the lower half of the ring is white, while the top half is blue, allowing the wearer to easily differentiate between day and night.

JACOB & CO.

EPIC SF24 – REF. 500.100.20.NS.PY.1NS

This 45mm titanium timepiece vertically indicates a secondary time zone with an unprecedented dual split-flap system that evokes the dynamic boards of the world's most famous travel terminals. Displaying the chosen city and the appropriate hour at the top of an openworked dial with micro-blasted gray center, upon which are indicated the hours, minutes and seconds, the instantaneous 24-hour world-time mechanism provides a fascinating vintage contrast to the otherwise ultra-modern design codes of the timepiece. Boasting a 48-hour power reserve, the watch's exclusive JCAA02 self-winding caliber is finished with Côtes de Genève and circular graining.

RICHARD MILLE

RM 11-02 AUTOMATIC FLYBACK CHRONOGRAPH DUAL TIME ZONE

In addition to the functions of the flyback chronograph, annual calendar, 60-minute countdown timer and 24-hour totalizer, the RM 11-02 Automatic Flyback Chronograph Dual Time Zone adds a second time zone. The PVD-treated titanium movement is fitted with two winding barrels and an automatic-winding rotor supplying circa 50 hours of power reserve. State-of-the-art, CNC 5-axis machines working with an accuracy of a few microns are a necessity for the creation of the multiple curves and arcs established in Richard Mille's case design aesthetic.

VACHERON CONSTANTIN

OVERSEAS DUAL TIME AUTOMATIC – REF. 47450/B01R-9404

Dressed in a case and bracelet made of radiant 18K rose gold, the 1222 SC automatic movement beats in style at a rate of 28,800 vph. A pink-gold stripe from 8:00 to 11:00 ruptures the dial's fine finishing to indicate the remainder of the watch's 40-hour power reserve. Enhanced with luminous materials on the hours and minutes, the three-hand time display is joined by a date subdial at 2:00, and secondary time zone with black and white day/night indicator. On the watch's solid caseback, a stamped medallion of the Overseas sailboat tells the story of the collection's heritage.

VACHERON CONSTANTIN

PATRIMONY TRADITIONNELLE WORLD TIME – REF. 86060/000R-9640

Combining Vacheron Constantin's rich multiple time zones heritage with a spirit of innovation, the self-winding 2460 WT caliber takes into account the modern reality of the world's 37 time zones. By aligning the reference city with a black triangle at 6:00, the owner has immediate access to time in all of the globe's regions, including those with partial time zones of half or even quarter hours. Brilliantly accomplished within a 42.5mm 18K pink-gold case, this masterpiece ingeniously uses three dials to obtain a supreme degree of complexity with optimal legibility. In addition to driving the central hours, minutes, and seconds, the movement precisely rotates a sapphire dial with 24-hour chapter ring and map-overlaying day/night indicator. Simplifying the use of the watch, all indication adjustments are made via a single crown. The Patrimony Traditionnelle World Time features a sapphire crystal openworked caseback.

ZENITH

PILOT MONTRE D'AERONEF TYPE 20 GMT – REF. 03.2430.693/21.C723

This self-winding tribute to Zenith's aviation history uses the generous size of its 48mm stainless steel case to display the hours, minutes, small seconds, and peripheral secondary time zone with absolute clarity and nighttime legibility. On the matte black dial enhanced with Arabic numerals made entirely of SuperLumiNova, the Elite 693 caliber, with 50-hour power reserve, animates a red-tipped hand that indicates the time in a chosen location via a peripheral 24-hour scale. On the caseback, the Zenith Flying Instruments logo pays tribute to the brand's proud aviation history.

ZENITH

PILOT TYPE 20 GMT 1903 – REF. 96.2431.693/21.C738

A tribute to the Wright brothers' contributions to aviation, the Pilot Type 20 GMT 1903 honors the first controlled, powered, heavier-than-air flight, achieved by Orville Wright in 1903, in a limited edition of 1903 pieces. The matte black dial is equipped with solid SuperLumiNova hour markers in SLN Old Radium. The hours, minutes, small seconds at 9:00 and a 24-hour secondary time zone with red-tipped hand play off the Arabic numerals on the dial's periphery. Worn on a vintage-style sturdy beige suede wristband that envelops the circumference of the DLC-coated titanium case, the watch boasts a caseback decoration of Orville Wright aboard his famous "Flyer." The dual time zone timepiece is powered by the 26-jewel self-winding Elite 693 caliber, oscillating at a frequency of 28,800 vph.

Calendars

MON TUE WED THU FRI SAT SUN
SEP OCT NOV DEC JAN FEB MAR APR MAY JUN JUL AUG

Piaget
Piaget Emperador Coussin Perpetual Calendar

Calendar Watches

In order to reproduce the particularities of the Gregorian calendar (with months of 28, 29, 30 or 31 days and the leap-year system), watchmakers have come up with feats of ingenuity. The vast palette of calendar watches extends from the simple calendar (requiring five corrections per year), to a plain day/date indication, annual calendars (one correction per year), perpetual calendars (accurate to the year 2100), and right the way through to the secular calendar (accurate just about forever). The more complicated the mechanisms within, the easier these timepieces are to use and to set. Traditional aperture-style designs are giving way to more freewheeling interpretations, with highly dynamic dials, retrograde hands, roller systems or completely transparent dial architectures.

THE VERY FIRST MEDIEVAL MONUMENTAL CLOCKS,

particularly the 14th- and 15th-century cathedral clocks already displayed a number of astronomical indications, some of which were related to the calendar.

Calendar displays appeared at a relatively early stage on pocket watches—well before the minutes and seconds hands. From the late 16th century onwards, certain astronomical watches showed the time, the date, the day the week, the month (with its length). Calendar watches subsequently proved a great success, but they were nonetheless merely "simple calendars" requiring manual adjustment after each month with less than 31 days, meaning five times a year. It was not until the late 18th century that the so-called "perpetual calendars" were developed, taking account of the variable lengths of the months and of the leap-year cycle.

The 1920s saw the emergence of the first perpetual calendar wristwatches representing authentic feats of miniaturization. In 1945, Rolex claimed the launch of the first wristwatch with date window, the Datejust, which a few years later was fitted with a magnifying glass to facilitate reading of the date. The date window subsequently became a classic feature of "simple" watches. In the 1980s, with the rebirth of mechanical watchmaking, the perpetual calendar made a spectacular comeback as a symbol of technical mastery and expertise, and all the major brands were keen to perfect this already highly sophisticated mechanism. The start of the new millennium has witnessed the breakthrough of a complication that was curiously rare until then, the annual calendar, which is more practical than the simple calendar, but less complex and less expensive than the famous perpetual calendars.

HOW DOES IT WORK?

The Gregorian calendar poses several challenges to watchmakers, particularly the varying lengths of the months and the periodical return of February 29. There are several different kinds of mechanisms, with distinct degrees of complexity.

SIMPLE CALENDAR

This is the name given to a calendar systematically indicating 31 days per month. A simple calendar must be manually adjusted on the first day of months following a month with fewer than 31 days, meaning five times a year (March 1st, May 1st, July 1st, October 1st and December 1st).

The simple calendar mechanism operates by means of a reducing gear train activated via the hour wheel. It moves one notch forward with every two turns of the hour wheel (24 hours). The display appears either on a disc bearing the date, or via a hand fixed to the 31-day star.

COMPLETE CALENDAR

A complete calendar looks very much like a perpetual calendar, in that it displays the date, the day, the month and generally the moonphase. But it does not take account of the varying lengths of the months and must therefore be corrected five times a year.

ANNUAL CALENDAR

The annual calendar automatically recognizes all months of 30 and 31 days, but does not take account of the duration of the month of February. It must therefore be corrected once a year, on March 1st. Annual calendars generally operate with the help of a wheel and pinion system of the same basic type as that driving the simple calendar, but considerably more complex.

PERPETUAL CALENDAR

The perpetual calendar is a highly sophisticated mechanism that not only recognizes months of 30 and 31 days, but also the 28 days of February and the quadrennial return of February 29th. It therefore possesses a mechanical "memory" of 1,461 days and theoretically requires no manual intervention. It does not however take account of the Gregorian correction relating to secular or century years and must be corrected for every secular year of which the first two figures are not divisible by 4—as will for instance be the case in 2100.

The perpetual calendar is distinguished from other types of calendar by the presence of a highly sophisticated toggle (or multiple-lever) system that pivots on its axis to drive the various displays. The indications may be displayed by discs or hands.

SECULAR OR CENTURY CALENDAR

The secular or century calendar takes account of the Gregorian correction made to years ending with 00. This extremely complex and rare calendar appears only on a handful of ultra-complicated watches (see the chapter on Multi-Complications).

Simple Calendars

From watches with a date window to those featuring calendars complete with day, date and month indications, there is a wide spectrum of simple calendar watches, which require correction at the end of each month lasting fewer than 31 days (i.e. five times a year). From a classic style to contemporary reinterpretation, watchmakers are giving free rein to their creativity, with an appealing selection of innovative designs, handed displays and retrograde indications—enough to satisfy any desire.

Rolex, *Oyster Perpetual Day-Date II*
A worthy heir to the first wristwatch to spell out the entire name of the day (launched in 1956), the Rolex Oyster Perpetual Day-Date II offers a choice of 26 languages for this indication. The date aperture is topped by the famous “Cyclops” lens, which magnifies the numeral 2.5x.

Parmigiani, *Ovale Pantographe*
Immediately recognizable by its highly original telescopic hour and minute hands, the Ovale Pantographe by Parmigiani displays the date through an opening shaped like the arc of a circle at 6 o’clock. It has an 8-day (192-hour) power reserve.

Movado, *Red Label Museum Calendomatic*
Inspired by a model launched in 1946, the Red Label Museum Calendomatic by Movado combines the famous understated black Museum dial with an original date display by means of a rotating disc bearing red Roman or Arabic numerals.

Patek Philippe
Gondolo 8 Days, Day & Date Indication, Ref. 5200

Endowed with an eight-day power reserve, the rectangular, Art Deco-inspired wristwatch Gondolo 8 Days, Day & Date Indication, Ref. 5200 from Patek Philippe displays instantaneous date and day displays, respectively with a pointer and through a large aperture.

Rolex, *Cellini Date*

Launched in 2014, Rolex's Cellini Date opts for a classic, refined aesthetic, with a sunray guilloché dial featuring a date subdial at 3 o'clock. The case, in white or rose gold, houses an automatic COSC-certified chronometer movement.

Jaquet Droz
Grande Seconde Quantième

On the Grande Seconde Quantième from Jaquet Droz, the date display is effected with a hand on the oversized seconds subdial at 6 o'clock, while hours and minutes share the off-centered subdial at 12 o'clock. The blue or brown dial is adorned with a Côtes de Genève motif.

Hamilton, *Flintridge Gent*

Inspired by a 1930s watch designed for athletes, the Flintridge Gent from Hamilton is equipped with a protective cover featuring two small apertures that reveal only the day and date.

Louis Erard, *Emotion*

On the Emotion model by Louis Erard, the date plays the starring role at 12 o'clock thanks to a disc that appears through an aperture in the guilloché dial.

Oris, *Big Crown Pointer Date*

A classic model from the Oris brand, the Big Crown Pointer Date displays the date using a central pointer, tipped with a half-moon to highlight a number around the periphery of the dial.

Meistersinger, *Pangea Day Date*

Equipped with a single hand for the hours and minutes, the Pangea Day Date by the German brand Meistersinger indicates the date and day at 12 o'clock by a red marker pointing to the numbers on rotating calendar rings.

Blancpain, *Retrograde Calendar*

The Retrograde Calendar from Blancpain's women's collection displays the date upon its mother-of-pearl dial using a sinuous hand tipped with a small star. It is also equipped with a moonphase and a flower-shaped oscillating weight.

Large Dates

Highly prized for their readability as well as their original touch, "large date" displays are becoming increasingly sought-after in recent years, whether appearing alone on the dial or associated with other complications. Several well-known brands have taken an interest in this ingenious mechanism, improving both its running and its reliability.

A. Lange & Söhne, *Grande Lange 1 Lumen*
Large date displays are closely associated with A. Lange & Söhne. A specialist in this type of display, the German brand uses its patented system on several models, including the Grande Lange 1 Lumen with a luminescent coating that enables date read-off even in the dark.

Jaeger-LeCoultre, *Rendez-Vous Perpetual Calendar*
Dedicated to feminine wrists, the Rendez-Vous Perpetual Calendar by Jaeger-LeCoultre displays the date, day and month on three small subdials. The complete four-digit date appears at 7:30 on the guilloché dial background.

Glashütte Original, *PanoMaticInverse*

The large date on the PanoMaticInverse, from Glashütte Original, appears in a double aperture placed off-center at 10 o'clock. The dial reveals the movement's bottom plate, as well as its regulating organ featuring a swan-neck fine adjustment system.

Zeitwinkel, *273°*

On the 273° model from Zeitwinkel, the patented large date display occupies a double aperture at 10:30. The refined dial design includes a power reserve display (72 hours) at 1:30 and a small seconds display at 6 o'clock.

Audemars Piguet
Tourbillon Grande Date
Edward Piguet

Combining two superimposed discs and an ingenious system of partial gears, the display created by Audemars Piguet is intended to be the largest on the market. Witness this ultra-sophisticated Edward Piguet Tourbillon Grande Date model.

Cecil Purnell, *Grande Date*

The Grande Date from Cecil Purnell uses transparent effects, revealing the two discs of the oversized date display, as well as part of the movement and the tourbillon carriage.

Perpetual Calendars: Classic Designs

Representing the ultimate form of calendar watch, the famous perpetual calendars are characterized by their multiple displays (date, day, month, leap-year cycle), generally enriched by moonphases. In their most classic version, they generally feature three or even four subdials arranged symmetrically in relation to the dial enter. But this restrained elegance sometimes conceals technical developments that considerably enhance the user-friendliness and the security of these ultra-sophisticated mechanisms.

Patek Philippe
Ladies First Perpetual Calendar, Ref. 7140
With its Ladies First Perpetual Calendar in rose gold, Patek Philippe presents its ultra-thin perpetual calendar—a classic from the manufacture—in a women's version for the first time.

A. Favre & Fils
Phoenix 10.1 Quantième à Grand Affichage Rotatif
On the Phoenix 10.1 Quantième à Grand Affichage Rotatif by A. Favre & Fils, the units numerals remain immobile, while a red arrow (for the dates from 1 to 9) or the 1, 2 or 3 o'clock numeral move to indicate the date.

Blancpain, *Perpetual Calendar with Under-Lug Correctors*
On its Perpetual Calendar with Under-Lug Correctors, Blancpain has developed an original system serving to integrate the correctors within the four bracelet lugs, on the wrist side. A press of the finger is enough to correct the date, day, month, leap-year cycle or moonphase display.

Blancpain

Villeret Traditional Chinese Calendar

To appeal to Chinese watch collectors, Blancpain has released the first wristwatch equipped with a traditional Chinese (lunisolar) calendar, with a "double-hour" indication, Zodiac signs, Chinese month and date, the five elements, the ten "celestial trunks" and leap months.

Audemars Piguet

Jules Audemars Perpetual Calendar

Equipped with an ultra-thin self-winding caliber, the Jules Audemars Perpetual Calendar model by Audemars Piguet is a perfect example of a classic and yet timeless design. The leap-year cycle appears in the center of the month subdial.

Perpetual Calendars: New Configurations

The design of perpetual calendars enables brands to play with various aesthetic arrangements in order to make their models stand out from traditional dials. The trend is towards a number of apertures placed either in a straight line or around the arc of a circle. Watchmakers also enjoy combining pointer- and window-type displays in sometimes highly original compositions, a sophisticated aesthetic approach that is associated with a number of technical accomplishments.

Cartier

Rotonde de Cartier Astrocalendaire

The Rotonde de Cartier Astrocalendaire shakes up conventional perpetual calendar arrangements with three concentric "tiers," representing day, month and date and circumnavigated by framed apertures. The leap year cycle appears on the caseback. All indications can be adjusted forwards or backwards using the crown.

IWC

Da Vinci Perpetual Calendar Digital Date and Month

The Da Vinci Perpetual Calendar Digital Date and Month by IWC innovates by displaying the date and month in two oversized instant-jump twin apertures at 9 o'clock and 3 o'clock. The crown can be used for fast calendar adjustments.

Buben & Zörweg

One Perpetual Calendar

In addition to the day and month appearing in two in-line apertures at 12 o'clock, the One Perpetual Calendar by Buben & Zörweg provides a large date display by means of two discs that are partially visible on the dial side. It is equipped with a silicon lever and escape wheel.

F.P. Journe, *Octa Quantième Perpétuel*

The Octa Quantième Perpétuel by F.P. Journe features a highly legible display by means of three instant-jump date/day/month apertures. The indications are adjusted by a three-position crown, apart from the fast month correction which has its own lever hidden between the strap lug at 1 o'clock.

Breguet, *Classique 7717BA*

On this self-winding Classique watch by Breguet, the pointer- or window-type displays are lined up along the same vertical axis, against the brand's characteristic guilloché dial.

Antoine Martin, *Quantième Perpétuel QP01*

Equipped with a silicon escapement that needs no lubrication, the Quantième Perpétuel QP01 from Antoine Martin stands out for its original aesthetic, with vertical displays for the day and month, large date and "new-style" guillochage.

Ulysse Nardin, *Perpetual Manufacture*

The Perpetual Manufacture by Ulysse Nardin stands out by its correction system enabling forward or backward adjustment of all calendar indications via the crown. It also displays a second time zone that can be adjusted using pushbuttons.

L. Leroy, *Osmior Retrograde Perpetual Calendar*

On its Osmior Retrograde Perpetual Calendar, the venerable French brand L. Leroy has opted for retrograde central hands to display the date and apertures for the day, month and leap-year indications.

A. Lange & Söhne
Lange 1 Tourbillon Perpetual Calendar Handwerkskunst

The Lange 1 Tourbillon Perpetual Calendar Handwerkskunst from A. Lange & Söhne presents its different indications in a very original way on an ornately hand-engraved dial: a small, off-center hours/minutes subdial, a large date display in a double aperture, the day with a retrograde hand at 9 o'clock and the month on a disc rotating around the dial circumference.

H. Moser & Cie
Endeavour Perpetual Calendar Black Edition

On its titanium Endeavour Perpetual Calendar Black Edition, H. Moser & Cie offers an unconventional, stripped-down conception of the perpetual calendar. The month is indicated by a small central hand that points to the appropriate hour marker—for example, 5 for the month of May. The calendar can be adjusted forwards or backwards at any time.

Perpetual Calendars: Retrograding and Roller Systems

How can one enliven a perpetual calendar model? By choosing a retrograde-type mechanism for one or several of the calendar indications, with a hand moving across the arc of a circle before jumping back to its point of departure (see the chapter on Alternative Displays). Another, much rarer, option is that of roller-type indications. This quest for originality paves the way for all manner of fanciful visual touches calling for a superlative degree of technical mastery.

Jean Dunand, *Shabaka*

The ultra-sophisticated Shabaka watch by Jean Dunand is equipped with a minute repeater striking on a cathedral chime, and is also distinguished by its roller-type display of the perpetual calendar indications, all of which jump instantly at midnight.

Maîtres du Temps, *Chapter Two*

The Chapter Two watch, created by Daniel Roth, Roger Dubuis and Peter Speake-Marin for the Maîtres du Temps brand, displays the months and days spelled out on two rollers—complete with two correctors housed on the caseback.

Jaquet Droz, *Perpetual Calendar*

Jaquet Droz, known for its consistently original designs, offers a Perpetual Calendar model with double retrograde display occupying the entire upper part of the dial, with days on the left and the date on the right.

Piaget, *FortyFive Perpetual Calendar*

On its FortyFive Perpetual Calendar, Piaget presents a dynamic configuration, with retrograde date at 3 o'clock, retrograde day at 9 o'clock and subdial displaying the month and leap-year cycle at 12 o'clock.

Parmigiani, *Toric Corrector*

Equipped with a minute repeater and a pusher enabling instant correction of all the calendar and moonphase functions, the Toric Corrector by Parmigiani Fleurier displays the retrograde date on the arc of a circle running from 8 to 4 o'clock.

Perpetual Calendars: The Transparent Approach

Visible mechanisms are all the rage in the watch industry at the moment. After introducing sapphire crystal casebacks on virtually all their complicated models, brands are vying with each other to find ingenious means of revealing all or part of the movement through the dial side. Perpetual calendars are no exception, and there is a vast range of models with openworked dials, skeletonized movements or transparent dials that now hide almost nothing of their inner mysteries, at least in visual terms. These spectacular stagings are often matched by some exceptional technical feats accomplished behind the scenes.

Parmigiani
Tonda 42 Retrograde Perpetual Calendar
On the Tonda 42 Retrograde Perpetual Calendar by Parmigiani Fleurier, the abbreviations of the day and month appear on two discs featuring original shapes and visible through the transparent dial.

Vacheron Constantin
Malte Retrograde Perpetual Calendar Openface
Vacheron Constantin has equipped its Malte Retrograde Perpetual Calendar Openface watch in platinum with a double sapphire crystal. Transparent subdials with white rotating "pallets" serve to indicate the day, month and leap-year cycle painted in black on one of the sapphire crystals.

Jaeger-LeCoultre
Master Eight Days Perpetual SQ
The Master Eight Days Perpetual from Jaeger-LeCoultre, endowed with an eight-day power reserve, is shown here in a dazzling skeletonized version featuring daringly curved bridges echoing the lines of longitude and latitude.

Patek Philippe
Grand Complication Ref. 5104
On the Grand Complication Ref. 5104 model by Patek Philippe, the days, months and the leap-year cycle number are printed directly on the sapphire crystal dial and dark "pallets" highlight the appropriate white transfer by contrast.

Cartier
Tortue Perpetual Calendar
Equipped with an in-house movement, the Tortue Perpetual Calendar by Cartier combines various types of display—including a central pointer-type date indication and a retrograde day running from 8 to 4 o'clock—appearing on an openworked dial.

Annual Calendars

Less complex than a perpetual calendar, the annual calendar enables brands to create watches that are more affordable and simpler to operate, while providing a degree of user-friendliness far superior to simple calendars in that they require just one correction per year. The last few years have brought a flourishing spate of annual calendar models featuring a great variety of displays and designs, as well as occasionally associating this complication with other useful functions.

Hermès, *Dressage Annual Calendar*

The Dressage Annual Calendar watch by Hermès, with its exclusive Vaucher Manufacture Fleurier movement, is distinguished by its pointer-type 270° retrograde date display and its two visible day and month discs.

Zenith, *Captain Winsor Annual Calendar*

Equipped with the automatic-winding El Primero caliber, Zenith's Captain Winsor Annual Calendar associates the functions of a 1/10-of-a-second chronograph with an annual calendar boasting three apertures for the date, day and month.

Omega
Seamaster Aqua Terra 150M Annual Calendar
On the Seamaster Aqua Terra 150M Annual Calendar from Omega, the date (complemented by an aperture for the day as well) need only be adjusted once per year, on March 1. The case, available in red gold, steel or a two-tone version, houses an exclusive caliber with co-axial escapement.

Pierre DeRoche, *GrandCliff QA Power Reserve*
The GrandCliff QA Power Reserve model by Pierre DeRoche, equipped with a Dubois Dépraz movement, combines the annual calendar with a chronograph featuring retrograding 60-minute and 6-hour counters, a flyback function and a large date.

Patek Philippe, *Annual Calendar, Ref. 4936*
Launched in 1996, the patented Annual Calendar from Patek Philippe has become a bestseller for the brand. It has been interpreted in different versions, including the reference 4936 ladies' diamond-set model with mother-of-pearl dial.

Parmigiani

Tonda Retrograde Annual Calendar

On Parmigiani's Tonda Retrograde Annual Calendar, the date is displayed using a retrograde hand traversing an arc from 8 to 4 o'clock, while the day and month (as a number from 1-12) appear in two small apertures.

Rolex, *Oyster Perpetual Sky-Dweller*

Rolex's travel watch Oyster Perpetual Sky-Dweller pairs a dual time zone with an annual calendar. The month is displayed discreetly in 12 apertures arranged around the dial. Setting the time zones and calendar is done using the crown.

ochs und junior, *Anno*

The epitome of minimalism, the contemporary Anno from ochs und junior displays the calendar indications using three circular scales: date (1 to 31) around the edges of the dial, day (1 to 7) on the central axis, month (1 to 12) above the hour marker at 6 o'clock. The annual calendar mechanism designed by the famous Ludwig Oechslin has only three moving parts—as opposed to 40 in a conventional system.

Hysek, *Kilada Annual Calendar*

Combining rose gold and black rubber in a highly contemporary design, the Kilada Annual Calendar possesses an annual calendar with a month counter at 8 o'clock and a large date at 12 o'clock.

Girard-Perregaux
Cat's Eye Annual and Zodiac Calendars

Designed for women, the Cat's Eye Annual and Zodiac Calendars watch combines a snail-shaped date display with a large comet's tail-shaped aperture revealing the ever-circling dance of the Zodiac signs.

AUDEMARS PIGUET

JULES AUDEMARS PERPETUAL CALENDAR – REF. 26390OR.OO.D093CR.01

The shiny brown dial on this Audemars Piguet model features counters that convey a complex mechanism in a legible and elegant fashion. Powered by a self-wound Manufacture caliber 2120/2802, the counters (month at 12:00, day of the week at 9:00, date at 3:00 and moonphase and leap year at 6:00) are all framed in 18K pink gold to match the applied hour markers and hands, as well as the 41mm case. Water resistant to 2atm, The Jules Audemars Perpetual Calendar rests on a brown alligator leather strap complete with an 18K pink-gold AP folding clasp.

BLANCPAIN

VILLERET QUANTIEME PERPETUEL 8 JOURS – REF. 6659-3631-55B

Blancpain's Villeret Quantième Perpétuel 8 Jours complements the collection's aesthetic—double-stepped case, white Grand Feu enamel dial, graceful hollowed-out central hands—with a beautiful, complete and legible perpetual calendar display. The 42mm 18-karat red-gold case houses the 32mm 5939A movement, which possesses a power reserve of 192 hours (eight days, hence the name) and powers, in addition to central hours and minutes, leap year cycle and month at 12:00, date at 3:00, small seconds and moonphase at 6:00, and day at 9:00. An alligator leather strap lined with alzavel completes the elegant ensemble.

CARL F. BUCHERER

PATRAVI CALENDAR – REF. 00.10629.08.33.01

Powered by the automatic-winding CFB A1004 caliber, the Patravi Calendar is complete with a patented dynamic shock absorber and patented central dual adjusting system, which requires only one adjustment. This model has a small seconds counter at 6:00 and indicates the day of the week at 9:00 and large date at 11:00. There is also a 53-week indicator in this ultra-legible model. The stainless steel case is complemented by twelve polished shiny hour markers on the sleek black dial.

DE BETHUNE

DB25 QP BLACK GOLD – REF. DB25QPARS8R

Combining De Bethune's signature spherical moonphase indication with a beautifully symmetrical perpetual calendar display, the DB25 QP Black Gold is housed in a 44mm rose-gold case featuring the fixed and hollowed lugs so characteristic of the brand. The black dial is decorated with hand-guilloché in 12 radiating sectors with apertures indicating the day of the week at 9:00 and the month at 3:00. A spherical moonphase indication in blackened zirconium and palladium at 12:00 spins against a backdrop De Bethune star-studded sky in black mirror-polished steel set with yellow-gold stars. A subdial at 6:00 indicates the date. The sapphire crystal caseback reveals the DB2324QP caliber, which boasts a silicon escape-wheel, self-regulating twin barrel, titanium/platinum balance spring with flat terminal curve and oscillating weight and a triple pare-chute shock-absorbing system—all De Bethune innovations.

F.P. JOURNE
QUANTIEME PERPETUEL

The Quantième Perpétuel presents a distinctive aesthetic through an uncluttered dial with apertures for the day, date and month. The instantaneous jump is reinforced by an ingenious F.P. Journe system designed to accumulate energy, release it instantaneously when the date, day or month changes, then slow it down at the end of its route. The varying lengths of the months are automatically taken into account, and the leap-year cycle is displayed at the center of the dial. The automatic-winding movement, constructed on the basis of the exclusive Octa calibre 1300.3, is manufactured in 18K rose gold and possesses 120 hours of power reserve. The Quantième Perpétuel is available in platinum or in 18K red gold, with 40mm or 42mm diameter.

GLASHÜTTE ORIGINAL

SENATOR PERPETUAL CALENDAR – REF. 100-02-22-05-05

Made by hand in the Glashütte manufacture, the Senator Perpetual Calendar exemplifies the ongoing quest for the perfect union of mechanical complications and art. All Senator Perpetual Calendar displays are arranged to form a clear, harmonious balance. The day, month and Glashütte Original Panorama Date display black numerals against a white background, and the warm silver dial shows a silver moonphase display on a blue disc. A discreet, unambiguous leap-year display presents a red dot to indicate a leap year, followed by yellow, black and white dots for successive years. The heart of the Senator Perpetual Calendar, visible through the sapphire crystal caseback, is the Glashütte Original Caliber 100-02, a 28,800 vph automatic movement featuring a screw balance with 18 weighted golden screws and a power reserve of more than 55 hours.

GLASHÜTTE ORIGINAL

SENATOR PERPETUAL CALENDAR – REF. 100-02-25-12-05

On this stainless steel, black-dial version of the Senator Perpetual Calendar, the moonphase appears on a black disc adorned with silver stars and a radiant silver moon. The timepiece's reset mechanism allows easy synchronization of the second hand with a time standard. Unlike other reset mechanisms, the second hand is connected neither to the winding stem nor the crown. Consequently, when the crown is pulled out, the balance remains in oscillation and the movement continues to run, significantly reducing material stress. Pressing a separate button on the side of the case at 8:00 activates the reset mechanism. Exquisitely finished components enhance the artistry of the Glashütte Original Caliber 100-02's design: the signature Glashütte three-quarter plate with Glashütte ribbing, skeletonized rotor with 21K gold oscillation weight, twin spring barrels and swan-neck fine adjustment.

GUY ELLIA

TIME SPACE QUANTIEME PERPETUEL

The Time Space, original ambassador of Guy Ellia, is transformed into a perpetual calendar for this version. A mechanical hand-winding Frédéric Piguet caliber harbored in a 46.8mm black-gold case allows the owner to read the hour, minute, day, month, date and moonphases, while taking leap years into account. With a brushed bottom plate of 7.75mm, this watch symbolizes elegance, complexity and thinness. A specific automatic winder is integrated to the watch box. This model is also available in white or pink gold.

JAEGER-LECOULTRE

MASTER GRANDE TRADITION A QUANTIEME PERPETUEL 8 JOURS SQ – REF. 50635SQ

The Master Grande Tradition à Quantième Perpétuel 8 Jours SQ watch demonstrates all Jaeger-LeCoultre's mastery of rare crafts such as skeletonizing, engraving and enameling. This watch's skeleton movement is cut away and decorated entirely by hand. In order to enhance the dial, its rim is decorated with guilloché work and covered with translucent blue enamel. A truly exceptional timepiece, it is also the only perpetual calendar watch with an 8-day power reserve. Its perpetual calendar functions are mechanically programmed until the year 2100, and are all activated by a single date corrector.

JAQUET DROZ

THE ECLIPSE MOTHER OF PEARL – REF. J012614570

The automatic-winding Jaquet Droz 6553L2 caliber powers The Eclipse Mother-Of-Pearl and its retrograde moonphase and calendar displays. The movement is complete with a 22K white-gold oscillating weight in addition to a double barrel. On the dial, the central hour and minute hands are accompanied by the date (indicated by hand) and day and month in an aperture at 12:00. The retrograde moonphase is at 6:00. All of these functions take place upon a white mother-of-pearl face. The ethereal appeal of this model is enhanced by eight red-gold star appliqués with rhodium treatment. The 18K white-gold case measures 39mm in diameter and is set with 248 diamonds totaling 1.62 carats in weight.

PATEK PHILIPPE

ANNUAL CALENDAR CHRONOGRAPH – REF. 5960/1A

Patek Philippe's new Ref. 5960/1A Annual Calendar Chronograph emanates the cool sheen of stainless steel. The new stainless steel case imbues the coveted Annual Calendar Chronograph with authoritative technical finesse that melds classic elegance with masculine sportiness. The hands and the calendar displays are powered by the Patek Philippe CH 28-520 IRM QA 24H caliber with flyback chronograph and annual calendar as well as power reserve and day/night indications. A heavy central rotor in 21K gold winds the watch automatically. Endowed with the Patek Philippe Hallmark, the movement combines the classic column-wheel control concept with a modern vertical disk clutch that eliminates the risk of hand jump or recoil when the stopwatch function is activated.

PATEK PHILIPPE

MEN GRAND COMPLICATIONS – REF. 5140R-011

This elegant model from Patek Philippe, affixed with the Patek Philippe Hallmark, is housed in a 37.2mm rose-gold case, with a silvery opaline dial that features baton rose-gold hour markers and two dauphine hands for the hours and minutes. The perpetual calendar has a mechanical memory spanning 1,461 days and needs no adjustment until the year 2100. The highly precise moonphase, integrated into the date dial at 6:00, will have a discrepancy of just one day after 122 years of operation. All calendar indications can easily be corrected via four pushbuttons recessed into the side of the case. Crafted entirely in the manufacture haute horology workshops, the self-winding 240 Q movement is composed of 275 parts yet boasts an ultra-thin 3.88mm profile.

PATEK PHILIPPE

MEN GRAND COMPLICATIONS – REF. 5940G-001

With a length of 44.6mm and a width of 37mm, this 18K white-gold Men Grand Complications watch from Patek Philippe features a very contemporary format with strong masculine appeal due to its modest height of just 8.6mm. Recipient of the Patek Philippe Hallmark, the caliber 240 Q movement is a masterpiece of technology crafted in its entirety in the manufacture's ateliers for complicated watches. The moonphase display deviates from the true position of the moon by one day every 122 years. The rate accuracy of the movement—responsible for timekeeping precision—ranges from -3 to +2 seconds per day. The model's timelessly elegant dial has a velvety, silvery white surface that perfectly matches the white-gold hue of the case. The time display is accompanied by three subsidiary dials for the calendar indications.

PIAGET

PIAGET EMPERADOR COUSSIN PERPETUAL CALENDAR – REF. G0A33019

The Piaget Emperador Coussin Perpetual Calendar is powered by the automatic-winding, ultra-thin Piaget 855P caliber. With a power reserve of 80 hours, this piece oscillates at 21,600 vph. The central hours and minutes display is accompanied by a small seconds counter at 4:00. Additionally, the perpetual calendar is complete with a retrograde date indicator at 3:00, a retrograde day indicator at 9:00, a month and leap year cycle indicator at 12:00 and a day/night indicator. This model is also equipped with a second time zone indicator at 8:00. All of these complications stand out on the blue dial that is framed by an 18K pink-gold case.

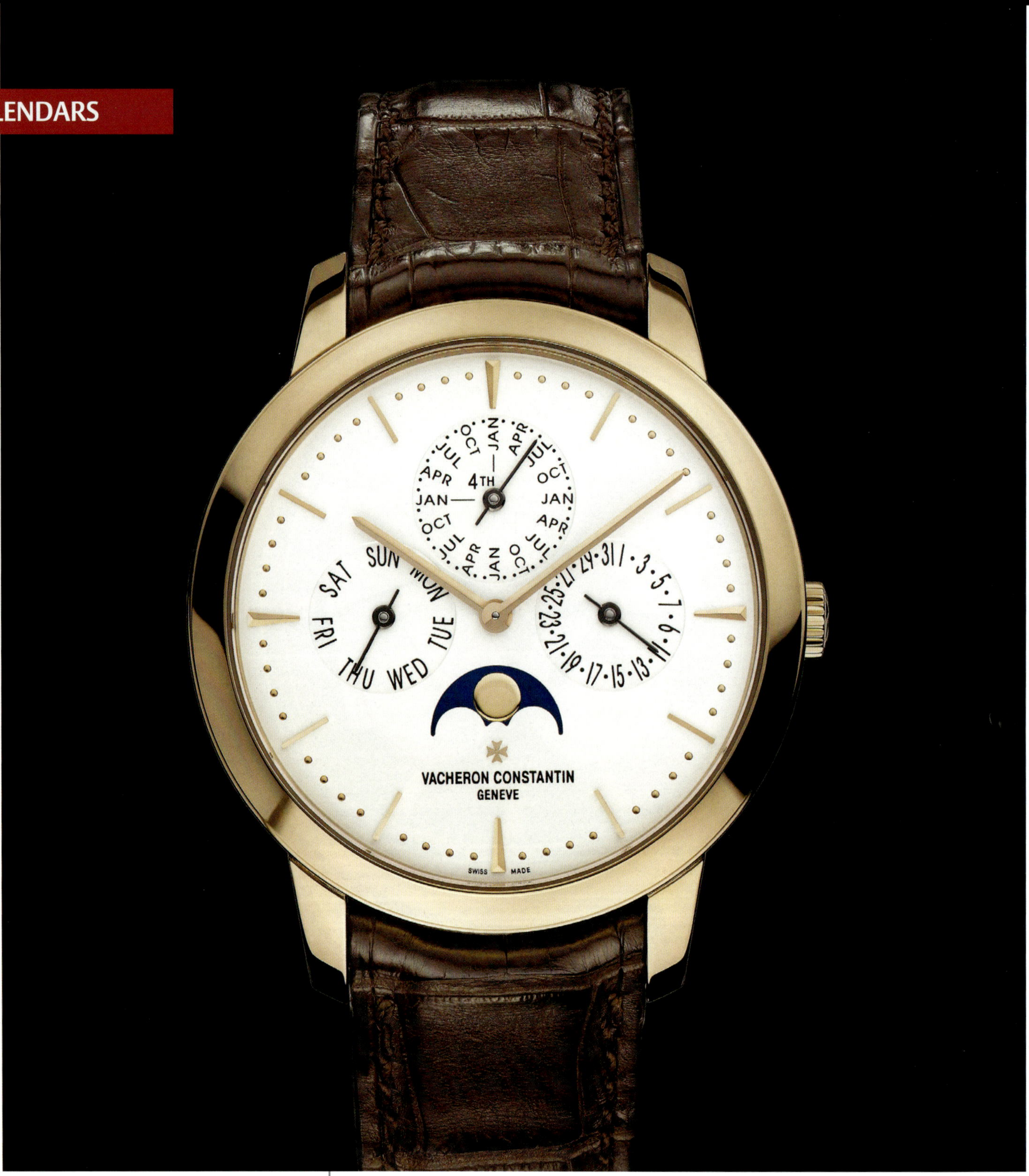

VACHERON CONSTANTIN

PATRIMONY CONTEMPORAINE PERPETUAL CALENDAR – REF. 43175/000R-9687

On its silver-toned opaline dial with baton- and triangle-shaped 18K pink-gold hour markers, this elegant timepiece presents a perpetual calendar so effortless that the moonphase at 6:00 seems to conduct the entire time orchestra. Via a month dial at 12:00, Vacheron Constantin provides an optimally unambiguous display of the leap-year indicator residing in its interior. Contained within the 41mm 18K pink-gold case, the Poinçon de Genève-honored 112 QP self-winding movement beats at a frequency of 19,800 vph, with a 40-hour power reserve. When the watch is turned over, the openworked back can be admired through a sapphire crystal.

ZENITH

CAPTAIN WINSOR ANNUAL CALENDAR – REF. 18.2070.4054/02.C711

With an annual calendar complication comprised of only nine mobile elements, as opposed to the traditional 30 to 40, this self-winding 18K rose-gold timepiece presents an ingenious construction of tremendous mechanical elegance. A pair of bold blue hands indicates the seconds and minutes of the column-wheel El Primero 4054 caliber. Three apertures indicate the date, day and month, crafting a complete calendar display that needs only one annual adjustment, when February turns to March. Finished with a silver-toned dial with central guilloché motif, the 42mm wristwatch boasts a frequency of 36,000 vph and power reserve of 50 hours.

Retrograde and Jumping Systems, Alternative Displays

Richard Mille
Dizzy Hands

Retrograde and Jumping

The circle described by the hands on a watch's dial is not the only way to display the time—alternative solutions do exist. Retrograde mechanisms, for example, use hands to swing across a semicircular arc before leaping back to their original positions, while jumping systems are equipped with numerals that appear through an aperture. Over the last few years, we have also seen the introduction of highly original display modes and time indications that boldly break with tradition. The time has now come for spectacular shows

Hautlence, *Avant-garde HLRQ03*

The Avant-Garde HLRQ03 model from Hautlence associates a jumping hour at 9 o'clock and a retrograde minute at 3 o'clock, linked by an exclusive connecting-rod/crank system—all housed within an open-worked architecture that allows for beautiful 3D effects.

Fonderie 47, *Inversion Principle Red Gold*

Featuring jumping hours, retrograde minutes, a central flying tourbillon rotating in three minutes and a six-day power reserve, the Inversion Principale by Fonderie 47 is also designed as a "charity watch" to raise funds for disarmament in Africa. It is available in a 10-piece limited edition.

MB&F, *Horological Machine N° 2*

The Horological Machine N° 2 by MB&F stands out for its resolutely original two-dial look. The right-hand dial displays jumping hours and concentric retrograde minutes, while the left-hand provides retrograde date as well as twin-hemisphere moonphase displays.

Spero Lucem, *La Jonction*

Paying tribute to a district of Geneva, the La Jonction model from the youthful Spero Lucem brand has a pure, restrained dial associating a jumping hour window at 11 o'clock with a retrograde minutes display at 1:30 and a flying tourbillon at 6 o'clock.

Dials go wild

Hands stopping at will, turning backwards or slowing down; hours and minutes indicated by tiny precious insects or by orbiting heavenly bodies; pivoting numerals, recomposed hour markers; watches with no time indication whatsoever... some watchmakers turn dials into a truly lively show in which visual attraction supersedes time read-off.

Chaumet
Attrape-moi... si tu m'aimes
The Attrape-moi... si tu m'aimes watch by the Parisian jeweler-watchmaker displays the hours and minutes in the form of a perpetual chase starring a spider and a bee (the brand emblem) rotating around the dial. The collection also features other insects such as the dragonfly and the butterfly.

Hermès, *Arceau Le temps suspendu*
On the Arceau Le temps suspendu watch by Hermès, a pusher serves to bring the hour and minute hands to a halt at about 12 o'clock, while a further press sets them running again as usual. Meanwhile, a mischievous hand whirls continually backwards on a tiny subdial with 24 graduations.

Pierre Kunz, *Infinity Looping*
On its Infinity Looping model, the Genevan watchmaker Pierre Kunz invented a spectacular new way of reading the time. Thanks to the gravitation of a satellite pinion that moves around a fixed rack, the hours and minutes whirl around the dial, tracing a rosette shape.

Christophe Claret, *Margot*

Un peu, beaucoup, passionnément, à la folie, pas du tout… With each press on the pushbu atch from Christophe Claret springs to life, randomly "plucking" one or two petals to come up with a romantic fortune ("He loves me… a little, a lot, passionately, madly, not at all…") that is displayed through an aperture at 4 o'clock.

Chanel

Mademoiselle Privé Lune et Comète

On the midnight-blue dial of Chanel's Mademoiselle Privé Lune et Comète, the hours are read off around the periphery using a moon-tipped hand, while a rounded comet counts out the minutes at the center of the dial.

Richard Mille

RM 63-01 Dizzy Hands

On the RM 63-01 Dizzy Hands from Richard Mille, one need only push the button in the center of the crown to make the dial rotate counterclockwise and the hour hand start spinning. Another push brings everything back to normal.

Ludovic Ballouard, *Half Time*

A creation from independent watchmaker Ludovic Ballouard, Half Time, the hour hand is replaced by a series of Roman numerals cut in half and thus illegible—apart from that of the hour in progress, instantaneously appearing at 12 o'clock. The minutes are shown by a retrograde hand.

RJ-Romain Jerome, *Day & Night*

The nine-piece Day & Night limited series by RJ-Romain Jerome is distinguished by its dial free of any temporal indication, other than two sequential tourbillons serving to tell day from night.

Linear displays

Some brands make a clean break from the circular sweep of dial hands or the segments featured on retrograde displays by offering linear horizontal or vertical displays of the hours, minutes or seconds, as well as for certain auxiliary indications (chronograph, power reserve, etc.). This approach is often coupled with the development of certain highly innovative mechanisms (cylinders, magnetic spheres, chains, belts, etc.).

Christophe Claret, *X-TREM-1*
On the X-TREM-1 from Christophe Claret, the retrograde hour and minute hands are replaced by two small steel spheres placed in two sapphire tubes on either side of the case and which move using magnetic fields.

Audemars Piguet, *Royal Oak Carbon Concept*

On its Royal Oak Carbon Concept watch, Audemars Piguet adopts a linear display for the 30-minute chronograph counter, composed of a double vertical scale appearing on the right-hand side of the dial (with tens from 1 to 2 and units from 0 to 9), while the elapsed time is indicated by the successive transitions of the various openings from white to black.

Urwerk

UR-CC1 Black Cobra

Inspired by the speedometers on certain 1950s cars, the UR-CC1 Black Cobra from Urwerk presents the linear display of the hours and minutes via two cylinders. Upon reaching the far end of its trajectory, the minute cylinder returns to its initial position in one tenth of a second, and this retrograde movement pushes the hour cylinder one unit forward.

Maurice Lacroix, *Masterpiece Seconde Mystérieuse*

The Masterpiece Seconde Mystérieuse by Maurice Lacroix offers a linear reading of the seconds in alternating vertical and horizontal 15-second cycles. The seconds hand appears to be floating or virtually levitating above a light-colored subdial.

RJ-Romain Jerome, *Spacecraft*

Designed by watchmakers Eric Giroud and Jean-Marc Wiederrecht, the Spacecraft by RJ-Romain Jerome boasts a striking design reminiscent of the spaceships featured in TV series from the 1970s and 80s. A disc indicates the minutes on the top of the case, while the retrograde linear jumping hour is displayed by a red cursor visible on the side of the watch.

Disc-type systems

These days, discs no longer hide beneath the dials. Orbiting like skies in the night sky, they create lively ballets, providing a stage for all manner of fanciful ideas.

RSW, *Outland*

On Outland from the brand RSW, the traditional hands make way for three concentric discs (hours, minutes and seconds) that move around a fixed point.

Bell & Ross, *BR01-92 Heading Indicator*

Inspired by the gyroscopic compass in aircraft cockpits, the dial of the BR01-92 Heading Indicator from Bell & Ross comprises three discs. The seconds are displayed in the center pointed to by a yellow marker and the minutes on an intermediate disc at 12 o'clock, while the hours appear opposite the yellow triangle on the outer disc.

Harry Winston, *Opus 7*

For Harry Winston, in the Opus 7, the watchmaker Andreas Strehler created an "alternating display." A first press on the crown guard turns the disc to bring the correct figure opposite the marker at 10 o'clock, a second press displays the minutes, and a third the power reserve.

Ressence, *Type 3*

The Type 3 model by Ressence is equipped with a satellite system with various rotating disc-type displays (hours, seconds, days), while the mobile dial itself indicates the minutes. The entire system is immersed in a visibility-enhancing liquid. The date appears around the periphery.

Roller-type systems

While still very rare, roller-type systems are sometimes used to indicate the hours and minutes, along with other temporal data. This touch of originality is in some cases associated with equally unusual designs.

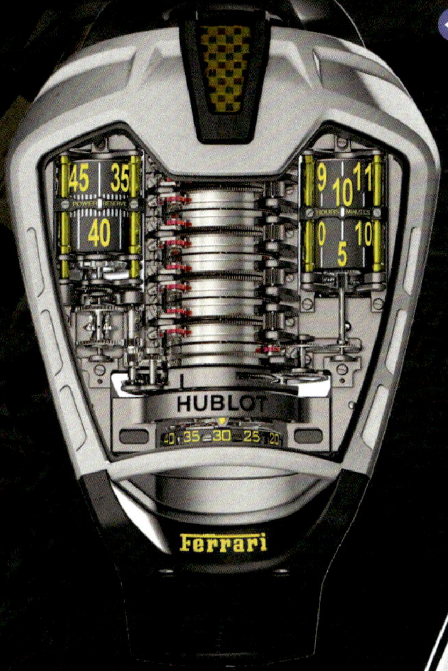

Hublot, *MP-05 LaFerrari*

A true watchmaking UFO, Hublot's MP-05 LaFerrari associates a highly original silhouette with a display of hours, minutes, seconds and power reserve on vertical cylinders. Its movement is equipped with an extraordinary power reserve of 50 days, using 11 barrels arranged in a line.

Cabestan, *Trapezium*

Equipped with a vertical tourbillon and a fusee-chain device, the highly original Cabestan Trapezium in titanium displays the hours, minutes, seconds and power reserve using a system of rotating drums.

Jean Dunand, *Shabaka*

The ultra-sophisticated Shabaka model by Jean Dunand is equipped with a cathedral-gong minute repeater mechanism and is also distinguished by its instant-jump roller-type perpetual calendar indications.

Maîtres du Temps, *Chapter Two*

The Chapter Two model, created by watchmakers Daniel Roth, Roger Dubuis and Peter Speake-Marin for the Maîtres du Temps brand, displays the months and days in full on two rollers controlled via two correctors recessed into the back of the case.

Rebellion, *T-1000 Gotham*

In addition to its imposing tank-like titanium case, and its impressive 40-day power reserve (ensured by six barrels), the T-1000 Gotham by Rebellion features a vertical indication of the hours and minutes appearing on two rollers.

3 D displays

In recent years, we have been witnessing an ever-increasing number of watches revolutionizing the indication and reading of time by means of pivoting, retractable, satellite-style or hydro-mechanical three-dimensional mechanisms—often combined with highly original case designs. These alternative systems are generally produced by the new watch brands or by independent designers and developers, sometimes working backstage for the major companies. While they do not always guarantee optimal legibility, they nonetheless display impressive inventiveness backed by peerless technical mastery.

Harry Winston, *Opus 12*

The Opus 12 from Harry Winston, developed in conjunction with Emmanuel Bouchet, proposes an unusual way to read time, with 12 pairs of hands that pivot to mark the hours and five-minute increments. A central retrograde hand counts off the five minutes between the increments.

MCT *Sequential Two – S200*

On the spectacular time display of the Sequential Two – S200 from MCT, the hour appears in turn in one of the four display zones on the dial by means of an original system of triangular prisms closely resembling animated advertising panels. At the turn of the hour, a C-shaped ring rotates 90° counterclockwise to reveal the following hour.

Bulgari, *Chronograph Papillon*

On the Bulgari Chronograph Papillon from the Daniel Roth Collection, the minutes are displayed by an ingenious central device that completes a 360° rotation in two hours and boasts two independent pivoting hands that take turns traversing a 180° segment that marks 0 to 60 seconds.

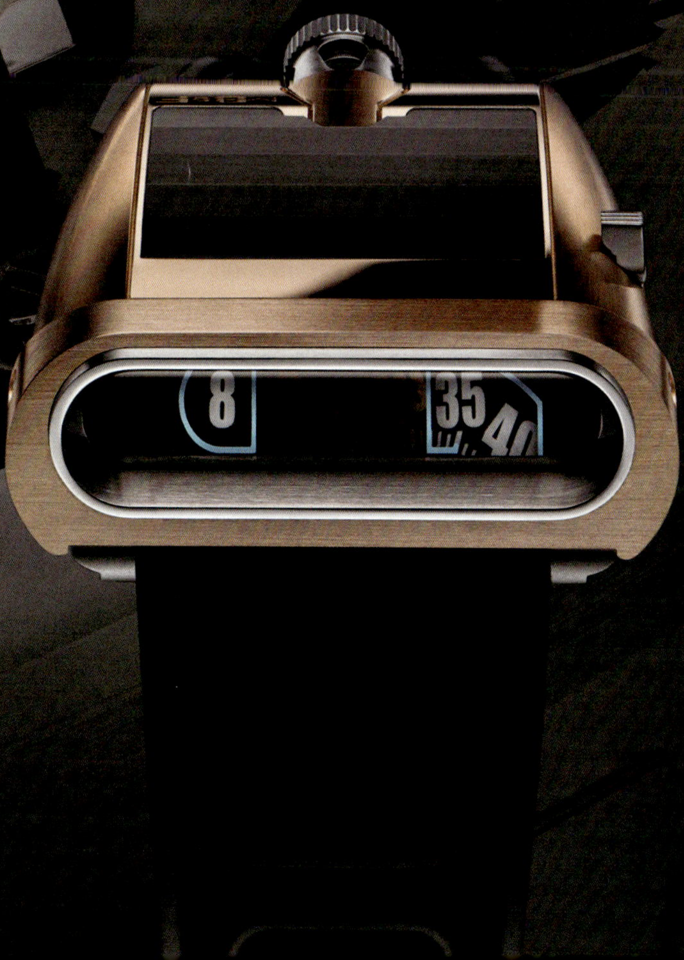

Urwerk, *UR-202*

The UR-202 by Urwerk indicates the time by means of three rotating cubes taking turns, while a telescopic hand displays the minutes by precisely following the angular path of the graduated scale.

MB&F, *Horological Machine N° 5 On the road again*

Paying tribute to supercars, the highly original Horological Machine N° 5 On the road again, by MB&F, features a system of louvers opening to let the light through and thus charge the SuperLumiNova of the hour and minute discs, which are placed flat on the movement and appear vertically on the side dial by means of an optical effect.

Cyrus, *Klepcys*

Created by the watchmaker Jean-François Mojon, the movement of the Cyrus Klepcys model displays the hour using a retrograde hand mounted upon two rotating cubes that indicate the day and night. The minutes and seconds are read off two central discs, and the date is indicated by a retrograde hand with a three-dimensional rotating tens column.

Harry Winston, *Opus 11*

Harry Winston's Opus 11, created by watchmaker Dennis Giguet, has no hands at all. Every 60 minutes, the large central numeral for the hour explodes into chaos before being reconstituted in a fraction of a second, using 24 mobile "pallets."

Harry Winston, *Opus 8*

For the Opus 8 by Harry Winston, Frédéric Garinaud devised a system of digital hours and minutes display inspired by pin art games. Activating a bolt on the side of the case causes the digits corresponding to the time to "rise up" from the dial, while the minutes are marked off on a vertical scale.

HYT, *H1 Red2*

The H1 Red2 by HYT (known as the "Hydro Mechanical Horologists") features a standout retrograde display of the hours by means of a fluorescent liquid propelled by the alternating compression/decompression motion of two mini-reservoirs with V-shaped pistons at 6 o'clock.

CHAUMET

ATTRAPE-MOI... SI TU M'AIMES CREATIVE COMPLICATION – REF. W16199-BC1

This seductive dance between a spider and Chaumet's iconic bee displays time with an imagination only matched by unprecedented mechanical innovation. Thanks to two off-centered rails on a brilliant-cut-diamond-set mother-of-pearl dial, the CP12V-XII self-winding caliber drives a display of time that eschews hands and conventional circularity. Pursuing independent paths upon an exquisitely crafted web, the pink-gold spider and diamond-studded bee indicate the hours and minutes, respectively. Framed by a bezel set with diamond indexes and 12 pink-gold cabochons, the two insects act out a wily game. A white leather strap set with two rows of diamonds ensures a luxurious fit around the wearer's wrist.

DE BETHUNE

DB28 DIGITALE – REF. DB28D

The DB28 Digitale, equipped with a mechanism powering multiple display modes—digital for the jumping hours, analog for the minutes and spherical for the moonphase—embodies a highly contemporary vision of a traditional 18th-century aesthetic. The central spherical moon indicates the moonphase with a degree of precision corresponding to a difference of just one lunar day every 1,112 years. The polished titanium case of the DB28 also defines the modern, even futuristic nature of the DB28 Digitale. The caseback reveals the finishing and technology of a Manufacture De Bethune movement. Several patents and innovations are displayed through this window on the DB2144 calibre, equipped with the exclusive De Bethune self-regulating twin barrel, circular silicon/white-gold balance wheel, balance-spring with flat end curve, as well as the triple pare-chute shock-absorbing system.

de GRISOGONO

MECCANICO – REF. MECCANICO DG N04

This ultra-modern timepiece in titanium and 18K pink gold is driven by a 651-component hand-wound caliber that uses sophistication in the service of a display of an intuitive and legible secondary time zone. Beneath the traditional circular indication of the time at 12:00, through which the wearer can appreciate the details of the movement, a digital display is animated by the instantaneous rotation of up to 12 individual micro-segments for each transition. This avant-garde module is composed of 23 cams, a set of gears and a trigger/synchronization device.

DIOR

CHIFFRE ROUGE C01 – REF. CD084C10A001

A timeless timepiece with a retrograde display completes the Chiffre Rouge collection and is the quintessential watch for the formal and stylish man. The delicate balance of the retrograde date display, day counter and power reserve indicator makes this model elegant in design while complex in mechanics. The Chiffre Rouge C01 timepiece is embellished with a cotton piqué patterned dial and sits gracefully on a hand-sewn alligator strap. It is also available in two versions: a silver opaline dial on a slate gray strap and a black dial on a black strap, both available in limited editions of 200 pieces each.

HUBLOT

BIG BANG UNICO BI-RETROGRADE CHRONO – REF. 412.OQ.1128.RX

Designed to commemorate the 2014 FIFA World Cup Brazil™, of which Hublot was the official timekeeper, the Big Bang Unico Bi-Retrograde Chrono is Hublot's first watch with a bi-retrograde display. When the chronograph is started, its second hand and minute hand move in an arc of 45 minutes (half a soccer match), with an area for 15 minutes' added time. Its Big Bang Unico case possesses a powerful 45.5mm diameter and its dial bears the national colors of Brazil—yellow and green—in reference to the host country of the World Cup. The chronograph, nicknamed "Soccer Bang," is available in two versions: an 18K King Gold version with carbon fiber bezel limited to just 100 pieces, and a black ceramic version, also with a carbon fiber bezel, which is limited to 200 numbered pieces.

JACOB & CO.

PALATIAL FLYING TOURBILLON JUMP HOUR RETROGRADE MINUTES REF. 150.510.24.NS.PB.1NS

Complete with jumping hours with an instantaneous-jump disk at 10:00 along with a retrograde minutes sector between 12:00 and 3:00, the Palatial Flying Tourbillon Jump Hour Retrograde Minutes is powered by a hand-wound JCBM02 caliber. With a power reserve of 100 hours, this model oscillates at 21,600 vph. The titanium case, with a diameter of 43mm, is topped off by a glimmering 18K rose-gold crown. The vibrant sea-blue hue of the mineral crystal dial contrasts with the gilt lettering and titanium tourbillon within. The power reserve is indicated on the caseback, where the movement's intricate finishes may be admired through a scratch-resistant sapphire crystal.

JAQUET DROZ

THE TWELVE CITIES AVENTURINE – REF. J010110270

The Twelve Cities Aventurine features a starry aventurine dial inlaid with a white mother-of-pearl center. The stainless steel case measures 39mm in diameter and the pushbutton at 2:00 powers the city selection. The minutes are centered while the hours jump through an aperture at 12:00. The watch is named for the aperture at 6:00, as 12 different cities can be shown there depending on which time zone the wearer is in. The self-wound Jaquet Droz 5153 caliber oscillates at 28,800 vph and has a power reserve of 68 hours. The rolled-edge handmade navy blue alligator leather strap is fastened with a stainless steel ardillon buckle.

PATEK PHILIPPE

MEN GRAND COMPLICATIONS – REF. 5496P-014

Dominated by a retrograde date display that delineates the border between the hour markers and the dial's center, this perpetual calendar from Patek Philippe also displays the day, month, moonphase and leap year cycle on its honey brown dial. The 39.5mm platinum case houses the automatic-winding 324 S QR caliber, which measures 28mm in diameter and 5.35mm in thickness, is comprised of 361 components and provides a power reserve of 35-45 hours. The Men Grand Complications Ref. 5496 is endowed with the Patek Philippe Hallmark as a guarantee of quality.

RICHARD MILLE

RM 63-01 DIZZY HANDS

When the pusher in the center of the crown of the RM 63-01 Dizzy Hands is pressed once, the sapphire glass dial begins to slowly rotate counterclockwise as the hour hand moves clockwise, both at different speeds. Press the pusher again, and everything on the dial returns to normal. The mechanism of the RM 63-01 Dizzy Hands, the caliber CRMA3, is a completely new movement fully developed by Richard Mille's engineers in Les Breuleux, a complex wonder, reinventing principles taken from chronograph design to achieve these artistic results, together with specially designed and researched parts that separate the hour and minute functions completely from one another. Designed in grade-5 titanium, the movement is housed in a new Richard Mille case with 18K red-gold front and back bezels and a central titanium caseband.

VACHERON CONSTANTIN

PATRIMONY CONTEMPORAINE RETROGRADE DAY AND DATE – REF. 86020/000G-9508

Powered by the mechanical automatic-winding Vacheron Constantin Caliber 2460 R31 R7, this Patrimony Bi-Retrograde is hallmarked with the Poinçon de Genève. Beating at 28,800 vph and offering 43 hours of power reserve, the 27-jeweled movement is equipped with a guilloché rotor. Retrograde day is provided at 6:00 and retrograde date is shown at 12:00. The three-piece case is polished and water resistant to 3atm. The sapphire caseback reveals the exquisite movement.

Diving

IWC
SCHAFFHAUSEN
AQUATIMER
DEEP THREE
ACT. DEPTH
MAX. DEPTH
25
15
SWISS MADE

IWC
Aquatimer Deep Three

Diving Watches

Diving watches have made spectacular progress since the first water resistant watch case appeared in 1926. The constant companions of adventurers and explorers of extreme environments, these instruments must possess a high level of water resistance, legibility and functionality—requirements scrupulously laid down by an official standard. Though their raison d'être is a resolutely technical one, diving watches have also been enthusiastically embraced by a wider public with a keen interest in daring feats and ultra-sporty looks. From models that are water resistant to 100m to high-performance models that descend thousands of meters below the waves, watchmakers offer a wide variety of diving watches, equipped with different security systems. There has been a recent resurgence in instruments featuring mechanical depth gauges, as well as models with an audible alarm (see chapter on Alarms). Makers of high-end diving watches also pay very close attention to aesthetics, offering models that combine performance under the sea and chic style on dry land.

WATER RESISTANCE, AS WELL AS LUBRICATION, HAS ALWAYS CONSTITUTED ONE OF HOROLOGERS' MAIN CONCERNS. This is not to say that all horological aficionados are budding Jacques Cousteaus.

However, water resistance represents a welcome protection, even on terra firma, against the two great enemies of watch movements: humidity and dust.

The first water resistant watch, the famous Oyster from Rolex, arrived on the scene in 1926, thanks to the invention of the "Screw Crown." In 1953, the brand created for Professor Auguste Piccard a watch that would accompany his bathyscaphe to 3,150m under the sea. In 1960, the Deep Sea Special would descend to over 10,916m down in the Mariana Trench in the Pacific Ocean, and ascended to the surface intact after having undergone pressure of more than a ton per square centimeter.

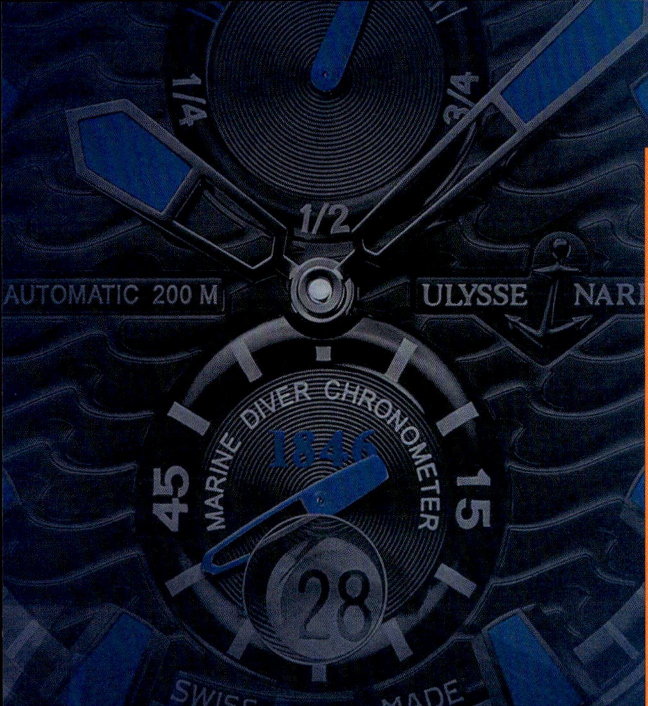

Besides explorers and scientists, the first diving watches often equipped the armed forces. This was notably the case with Panerai in the 1930s and 1940s, made for the Italian Navy, with the models' luminescent indications on the dial and a crown system guaranteeing water resistance to 200m.

In the 1950s, with the rise of water sports, came the first collections devoted to diving watches. Rolex set the tone in 1953 with its famous Submariner, water resistant to 100m (now 300m). That same year, Blancpain launched its Fifty Fathoms, created in collaboration with two "frogmen" from the French Army. This watch, initially resistant to 50 fathoms of depth—hence its name—stands out for its black dial, bezel and strap. It would later be released in countless other versions.

Ever since, diving watches have become the quintessential sporting watch, and horologers have continued to increase their resistance and robustness.

HOW DOES IT WORK?

The definition of an authentic "diving watch" conforms to an official Swiss industry standard, or "norme de l'industrie horlogère suisse" (NIHS 92-11/ISO 6425). To earn the title of diving watch, a timepiece must not just possess a water resistance superior to the average watch (minimum 100m), usually ensured by a screw-down crown and caseback. It must also offer all the functions that are indispensable to amateur or professional divers, notably a system that allows one to preset and check the time of the dive, with a device preventing any accidental manipulation. This is usually done by a turning bezel whose unidirectional rotation prevents the diver from inadvertently extending the remaining dive time.

The dial usually features large numerals and highly visible hands, generously covered in luminescent coating. The metal strap or rubber bracelet is often equipped with an extension system that stretches over the diving suit. The highest-performance models possess a valve that eliminates any helium within the watch. This highly volatile gas, used to facilitate breathing in pressurized chambers, gets inside the watch case very easily, but the case must be totally evacuated of helium when re-ascending to avoid the build-up of excessive pressure.

A SMALL REMINDER ABOUT WATER RESISTANCE

Contrary to what is all too often believed, the indication of water resistance in meters is a technical standard that does not correspond to an absolute value. The phrase "water resistant to 30m" does not imply that the watch can be worn to that depth under any conditions whatsoever. The movements of the wearer (dives, jumps, swimming, etc.) can increase the pressure considerably. For all aquatic activities, it is therefore best to stick to watches that possess a water resistance of at least 100m.

DIVING CHRONOGRAPHS

The inherently sporty nature of the two functions means that horologers often add chronographs to diving watches. However, it bears noting that most of these instruments, though perfectly "water resistant," cannot be activated in chronograph mode during the dive, or else water will seep into the case via the pushbuttons.

Re-Releasing Flagship Models

Several well-known brands are re-releasing or re-interpreting their emblematic diving watches, especially the pioneering models of the 1950s. Though their look might be similar to the original, their performances usually benefit from the enormous technical progress made in the intervening years.

Tudor, *Heritage Black Bay*

Inspired by a 1954 model, Tudor's Heritage Black Bay model (water resistant to 200m) stands out for its "vintage chic" design, with a convex dial, blue rotating bezel, "snowflake" hands and large luminescent hour markers.

Longines, *The Longines Heritage Diver*

Inspired by two models from the 1970s, The Longines Heritage Diver—available in a plain or chronograph version—stands out for its tonneau-shaped case, water resistance to 300m, and a bidirectional interior rotating bezel activated by the crown at 2 o'clock.

Blancpain, *Fifty Fathoms Bathyscaphe*

To mark the 60th anniversary of its historical Fifty Fathoms model (1953), which has now given rise to an entire collection of diver's watches and chronographs, Blancpain has created the Fifty Fathoms Bathyscaphe, inspired by the submarine of the Swiss adventurer Jacques Piccard.

Vulcain, *Nautical Seventies*

A tribute to a 1970s model, the Vulcain Nautical Seventies is distinguished by an alarm that is easily audible under water, water resistance to 300m and a decompression-stop indication system, which is adjusted using a screw-down crown at 4 o'clock.

Omega, *Seamaster PloProf 1200M*

Omega re-interpreted its 1970 Seamaster 600M model, also called PloProf (for "plongeurs professionnels," or professional divers), in a modernized version that is water resistant to 1200m, with a co-axial escapement and a security pushbutton that locks and unlocks the rotating bezel.

Rolex, *Oyster Perpetual Submariner*

Heir to the legendary 1953 model, the Rolex Oyster Perpetual Submariner is water resistant to 300m thanks to a patented Triplock screw-down crown system. The unidirectional rotating bezel is in ultra-resistant scratchproof ceramics.

Champions of Style

From sporty chic models destined mainly for poolside cocktails to professional instruments equipped for confronting the watery depths, diving watches exist in a vast array of choices. Steel occasionally gives way to gold or unusual materials—ceramics, high-tech alloys—and the quest for high performance is increasingly paired with close attention to style. These are watches that shine underwater, but also on dry land.

Eberhard & Co, *Scafodat 500*

The Scafodat 500 by Eberhard & Co features a system serving to extend the steel bracelet so as to ensure an easy fit over a diving suit, as well as a 4 o'clock crown activating the inner bezel ring with its dive-time indications.

Alpina

Extreme Diver 300 Orange

To ensure optimal readability in the ocean depths, the Extreme Diver 300 Orange model by Alpina is equipped with a black rotating bezel with orange indications and a minute hand (the most important one when diving) that is also orange-accented.

Gucci, *Gucci Dive*

Powered by a Girard-Perregaux movement, the Gucci Dive watch (water resistant to 300m) comes in steel or pink-gold versions, with a black PVD coating on both models. It features an eminently readable and elegant design.

Audemars Piguet, *Royal Oak Offshore Diver 42 mm*

The Royal Oak Offshore Diver 42 mm by Audemars Piguet (water resistant to 300m) has been interpreted in a stylish matte white high-tech ceramic version, with a strap in different colors of rubber. The rotating inner bezel ring, which calculates dive times, is adjusted via a crown at 10 o'clock.

Maurice Lacroix, *Pontos S Extreme*

On the Pontos S Extreme by Maurice Lacroix, crafted in an ultra-light alloy named Powerlite®, dive-time control is handled via a crown-driven rotating inner bezel ring.

Ulysse Nardin, *Marine Diver Black Sea / Blue Sea*

Water resistant to 200m, the Marine Diver Black Sea / Blue Sea by Ulysse Nardin stands out with its total-black look offset by blue hands and hour markers, as well as its steel case clad in vulcanized rubber. The dial and the unidirectional bezel are adorned with a wave pattern.

JEANRICHARD, *Aquascope Blue "Hokusai" Dial*

Water resistant to 300m, JEANRICHARD's Aquascope is available in several versions, including this model with a steel case, blue rubber strap and a dial adorned with a motif inspired by a print by the famous Japanese artist Katsushika Hokusai (1760-1849).

Cartier, *Calibre Diver*

The first diving watch from Cartier, the Calibre Diver (water resistant to 300m) plays off the brand's design codes, particularly its large Roman numerals. The unidirectional rotating bezel, with highly resistant black coating, has 120 notches enabling half-minute adjustments.

Super Divers

Certain companies have chosen to compete by developing exceptional instruments capable of descending to 1,000m, 3,000m, 6,000m or even 12,000m underwater! These super divers possess ultra-robust and innovative architectures, reinforced water resistance systems and even—for the highest performers—cases filled with liquid that is immune to pressure.

Bell & Ross, *Hydromax*

Equipped with a quartz movement, the Hydromax from Bell & Ross is water resistant to the amazing depth of 11,000m, thanks to its case, filled with an oily liquid that is immune to pressure.

Rolex, *Oyster Perpetual Deepsea*

Rolex has reaffirmed its sovereignty over the deep seas with its launch of the Oyster Perpetual Deepsea, water resistant to 3,900m. To resist such colossal pressures, this extraordinary instrument benefits from a patented new case architecture called Ringlock. The "D-Blue" dial is distinguished by its "ombre" dial, which goes from blue to deepest black.

Pita, *Oceana*

Using two patented mechanisms that enable time setting by rotating the base and "reinforced magnetic gears," the Oceana model from Barcelonan brand Pita—water resistant to 5,000m or 2,000m—can lay claim to being "the only watch that you can set underwater."

Rolex, *Oyster Perpetual Deepsea Challenge*

Water resistant to 12,000m, Rolex's experimental diving watch Oyster Perpetual Deepsea Challenge accompanied director and explorer James Cameron in 2012 to 10,898m underwater in the Mariana Trench, the deepest solo dive in history.

Breitling, *Superocean Chronograph M2000*

With a patented system of magnetic pushbuttons that operate through the case itself, the Superocean Chronograph M2000 from Breitling is presented as the only chronograph in the world that is watertight and functional at 2,000m underwater.

Girard-Perregaux, *Seahawk II Pro*

On its Seahawk II Pro, water resistant to 3,000m, Girard-Perregaux has shifted the crown to 4 o'clock to prevent it from getting in the way of the movements of the wrist, and equipped it with a large crown-guard that is both ergonomic and highly original.

Doxa, *SUB 4000T Professional*

Water resistant to 4,000 feet (1,220m), Doxa's SUB 4000T Professional stands out by its orange dial, helium valve, "Safe Dive" power reserve indicator and its asymmetrical case, shaped to protect the crown.

CX Swiss Military Watch™, *20 000 FEET*

The watchmaker CX Swiss Military Watch™ struck a blow for the record books with its 20 000 FEET, a mechanical timepiece that can plunge to more than 6,000m of depth.

Porsche Design, *P'6780 Diver*

On the P'6780 Diver from Porsche Design, water resistant to 1,000m, the automatic movement is lodged inside a pivoting chamber. When the chamber is locked, the rotating bezel and crown are immobilized. The user must unlock it to set the dive time.

Super Luminous

Diving watches must offer optimal readability in the extreme darkness prevailing at great depths. For this purpose, they are generally equipped with oversized numerals, indexes and hands that are treated with luminescent coating. Some brands have developed "lighting" systems offering even higher performance.

Ball Watch, *Engineer Hydrocarbon NEDU*
On Ball Watch's Engineer Hydrocarbon NEDU (water resistant to 600m), the hour markers and hands are equipped with micro-tubes filled with a luminescent gas. This Swiss technology guarantees a luminosity 100 times more intense than other methods.

Luminox, *Deep Dive*
The Luminox Light Technology System, in operation on the Deep Dive by Luminox, guarantees visibility in any conditions, thanks to its micro-tubes reflecting a blue light—the last color perceivable by the human eye after a certain depth.

Watches with Depth Gauges

Over the last few years, we have witnessed the return of a function that is not strictly horological, but very useful for divers: the depth gauge, or bathometer. It is a tempting alternative—or complement—to the electronic wristworn computers that are currently flooding the market.

IWC, *Aquatimer Deep Three*

The Aquatimer Deep Three from IWC (with a titanium case water resistant to 100m) is equipped with a mechanical depth gauge using two pointers in two different colors. The blue pointer indicates the current depth and the red one indicates the maximum depth achieved during the dive (up to 50m).

Blancpain, *X Fathoms*

Blancpain's X Fathoms houses a mechanical depth gauge that operates to 90m, with memory of the maximum depth achieved and a separate, ultra-precise (+/-30cm) display for the 0-15m zone, as well as a five-minute retrograde counter for the decompression stages.

RICHARD MILLE

RM 025 TOURBILLON CHRONOGRAPH DIVER'S WATCH

Within the RM 025 Tourbillon Chronograph Diver's Watch ticks the caliber RM 025, crafted in carbon nanofiber. The RM caliber 025 unites the complication of a tourbillon with the notorious complexities of a chronograph movement. The unique tripartite case, water resistant to 30atm, is complemented by a torque-limiting crown and a new design of watertight pushers. The bezel is constructed of three layers and turns unidirectionally in order to avoid timing miscalculations. The production and assembly of the RM 025 Tourbillon Diver's watch takes many months, with only a select number of watchmakers able to cope with its complex assembly and testing. The model is available in titanium, fitted with 18K red-gold lugs and a rubber diver's strap.

Brand Directory

A. LANGE & SÖHNE
15 Altenberger Strasse
01768 Glashütte
Germany
Tel: 49 35053 440
USA: 800 408 8147

AUDEMARS PIGUET
1348 Le Brassus
Switzerland
Tel: 41 21 845 14 00
USA: 212 758 8400

BELL & ROSS
8 Rue Copernic
75016 Paris
France
Tel: 33 1 73 73 93 00
USA: 888 307 7887

BLANCPAIN
6 Chemin de l'Etang
1094 Paudex
Switzerland
Tel: 41 21 796 36 36
USA: 877 520 1735

BREGUET
1344 L'Abbaye
Switzerland
Tel: 41 21 841 90 90
USA: 866 458 7488

CARL F. BUCHERER
1805 South Metro Parkway
Dayton, OH 45459
USA: 800 395 4306
Switzerland: 41 41 369 74 80

CHANEL
25 Place du Marché St. Honoré
75001 Paris
France
Tel: 33 1 40 98 50 00
USA: 212 688 5055

CHAUMET
12 Place Vendôme
75001 Paris
France
Tel: 33 1 44 77 24 00

CHRISTOPHE CLARET
2 Route du Soleil d'Or
2400 Le Locle
Switzerland
Tel: 41 32 933 80 80
USA: 954 610 2234

CORUM
Rue du Petit-Château 1
CH-2301 La Chaux-de-Fonds
Switzerland
Tel: 41 32 967 06 43
USA: 1 305 792 0884

DE BETHUNE
6 Granges-Jaccard
1454 La Chaux L'Auberson
Switzerland
Tel: 41 24 455 26 00
USA: 917 907 1127

DE GRISOGONO
39 Chemin du Champs des Filles
Bâtiment E
1228 Plan-les-Ouates
Switzerland
Tel: 41 22 817 81 00
USA: 212 439 4220

DIOR HORLOGERIE
44 Rue François 1er
75008 Paris
France
Tel: 33 1 40 73 59 84
USA: 212 931 2700

F.P. JOURNE
17 rue de l'Arquebuse
1204 Geneva
Switzerland
Tel: 41 22 322 09 09
USA: 305 572 9802

FRANCK MULLER
22 Route de Malagny
1294 Genthod
Switzerland
Tel: 41 22 959 88 88
France: 33 1 53 67 44 39
USA: 212 255 8499

FREDERIQUE CONSTANT SA
32 Chemin du Champ des Filles
1228 Plan-les-Ouates
Switzerland
Tel: 41 22 860 04 40
USA: 877 619 2824

GIRARD-PERREGAUX
1 Place Girardet
2301 La Chaux-de-Fonds
Switzerland
Tel: 41 32 911 33 33
France: 33 1 72 25 65 43
USA: 201 355 4523

GLASHÜTTE ORIGINAL
1 Altenberger Strasse
01768 Glashütte/Sachsen
Germany
Tel: 49 3 50 53 460
France: 33 1 53 81 22 68
USA: 201 271 1400

GUY ELLIA
16 Place Vendôme
75001 Paris
France
Tel: 33 1 53 30 25 25
USA: 212 888 0505

HUBLOT
33 Chemin de la Vuarpillière
1260 Nyon 2
Switzerland
Tel: 41 22 990 90 00
USA: 800 536 0636

IWC
15 Baumgartenstrasse
8201 Schaffhausen
Switzerland
Tel: 41 52 635 65 65
USA: 800 492 6755

JACOB & CO.
1 Chemin de Plein-Vent
1228 Arare
Switzerland
Tel: 41 22 310 69 62
USA: 212 719 5887

JAEGER-LECOULTRE
Rue de la Golisse 8
1347 Le Sentier
Switzerland
Tel: 41 21 845 02 02
USA: 212 308 2525

JAQUET DROZ SA
2 Allée du Tourbillon
2300 La Chaux-de-Fonds
Switzerland
Tel: 41 32 924 28 88
France: 33 1 53 81 22 00
USA: 201 271 1400

LONGINES
Saint-Imier 2610
Switzerland
Tel: 41 32 942 54 25
USA: 201 271 1400

PARMIGIANI
11 Rue du Temple
2114 Fleurier
Switzerland
Tel: 41 32 862 66 30
USA: 305 260 7770

PATEK PHILIPPE
141 Chemin du Pont du Centenaire
1228 Plan-les-Ouates
Switzerland
Tel: 41 22 884 20 20
USA: 212 218 1240

PERRELET
7 Rue Bubenberg
2502 Biel/Bienne
Switzerland
Tel: 41 32 346 26 26
USA: 954 575 7980

PIAGET
37 Chemin du Champ-des-Filles
1228 Plan-les-Ouates
Switzerland
Tel: 41 22 884 48 44
USA: 866 374 4430

RICHARD MILLE
11 rue du Jura
2345 Les Breuleux Jura
Switzerland
Tel: 41 32 959 43 53
USA: 310 205 5555

ROGER DUBUIS
1217 Meyrin 2 Geneva
Switzerland
Tel: 41 22 783 28 28
France: 33 1 58 18 14 67
USA: 888 738 2847

TAG HEUER
6A Louis-Joseph Chevrolet
2300 La Chaux-de-Fonds
Switzerland
Tel: 41 32 919 80 00
USA: 973 467 1890

VACHERON CONSTANTIN
10 Chemin du Tourbillon
1228 Plan-les-Ouates
Switzerland
Tel: 41 22 930 20 05
USA: 212 713 0707

VULCAIN SA
4 Chemin des Tourelles
2400 Le Locle
Switzerland
Tel: 41 32 930 80 10

ZENITH
2400 Le Locle
Switzerland
Tel: 41 32 930 62 62
USA: 973 467 1890